System-Wide Perspective for Life Cycle Assessment of CO_2-based C1-Chemicals

Ökologische Bewertung von CO_2-basierten C1-Chemikalien aus einer systemweiten Perspektive

Von der Fakultät für Maschinenwesen der
Rheinisch-Westfälischen Technischen Hochschule Aachen
zur Erlangung des akademischen Grades eines Doktors
der Ingenieurwissenschaften genehmigte Dissertation

vorgelegt von

André Dirk Sternberg

Berichter: Univ.-Prof. Dr.-Ing. André Bardow
Professeur Titulaire François Maréchal, Ph.D.

Tag der mündlichen Prüfung: 13.01.2017

Diese Dissertation ist auf den Internetseiten der Universitätsbibliothek online verfügbar.

Aachener Beiträge zur Technischen Thermodynamik Band 11
André Dirk Sternberg

System-Wide Perspective for Life Cycle Assessment of CO_2-based C1-Chemicals

Ökologische Bewertung von CO_2-basierten C1-Chemikalien aus einer systemweiten Perspektive

ISBN: 978-3-95886-193-0

Das Werk einschließlich seiner Teile ist urheberrechtlich geschützt. Jede Verwendung ist ohne die Zustimmung des Herausgebers außerhalb der engen Grenzen des Urhebergesetzes unzulässig und strafbar. Das gilt insbesondere für Vervielfältigungen, Übersetzungen, Mikroverfilmungen und die Einspeicherung und Verarbeitung in elektronischen Systemen.

Bibliografische Information der Deutschen Bibliothek
Die Deutsche Bibliothek verzeichnet diese Publikation in der Deutschen Nationalbibliografie; detaillierte bibliografische Daten sind im Internet über http://dnb.ddb.de abrufbar.

Herstellung & Vertrieb:

1. Auflage 2017
© Wissenschaftsverlag Mainz GmbH - Aachen
Süsterfeldstr. 83, 52072 Aachen
Tel. 0241/87 34 34
Fax 0241/87 55 77
www.Verlag-Mainz.de

ISSN: 2198-4832

Satz: nach Druckvorlage des Autors
Umschlaggestaltung: Druckerei Mainz

printed in Germany
D82 (Diss. RWTH Aachen University, 2017)

'The world has enough for everyone's need, but not enough for everyone's greed.'

Mahatma Gandhi

Vorwort

Die vorliegende Arbeit entstand am Lehrstuhl für Technische Thermodynamik der RWTH Aachen im Rahmen meiner Tätigkeit als wissenschaftlicher Mitarbeiter. Ein sehr großer Dank gebührt Prof. André Bardow für seine stetige Unterstützung. Neben der fachlichen Unterstützung denke ich da z.B. auch an seinen Einsatz beim Review-Prozess meines ersten Papers. Des Weiteren gilt mein Dank Prof. François Maréchal für die Übernahme des Koreferats im Promotionsverfahren. Seine Kommentare haben die Qualität dieser Arbeit nochmal gesteigert. Ich danke Prof. Alexander Mitsos für die Übernahme des Vorsitzes des Prüfungskomitees.

Ebenso möchte ich mich bei allen Mitarbeitern des LTT bedanken. Es war immer eine tolle Atmosphäre und ich habe sehr gern am LTT gearbeitet. Besonderer Dank gilt natürlich meiner Arbeitsgruppe – der E$T. Ihr habt mir die Entscheidung nach Aachen zu gehen sehr leicht gemacht und euch direkt an meinem ersten Tag in Aachen um mich "gekümmert". Für fachliche Diskussionen gilt mein Dank insbesondere der LCA-Gruppe der ersten Stunde: Johannes Jung, Niklas von der Assen und Arne Kätelhön. Niklas danke ich darüber hinaus für die schöne gemeinsame Zeit und inspirierende Arbeitsatmosphäre in unserem Büro. Des Weiteren gilt mein Dank den Studenten, die mich in meiner Arbeit am LTT unterstützt haben.

Zum Schluss möchte ich den wichtigsten Menschen in meinem Leben danken. Ich danke meinen Eltern für die Unterstützung in allen Lebenslagen. Ich danke Merle und Arne für die Freude die sie mir schenken in jeder Sekunde die ich mit ihnen verbringe. Ich danke Candy für alles.

Aachen, im Dezember 2017 *André Sternberg*

Contents

List of Figures

List of Tables

Notation

Abbreviations

ALCA	attributional life cycle assessment
BEV	battery electric vehicle
CAES	compressed air energy storage
CCS	carbon (dioxide) capture and storage
CHP	Combined heat and power
CLCA	consequential life cycle assessment
COP	coefficient of performance
CRF	capital recovery factor
DRM	dry reforming of methane
DSM	demand side management
EI	environmental impact
eq	equivalent
EU-27	European Union (in the period 2007–2013 when it had 27 countries)
FD	fossil depletion
FT	Fischer-Tropsch reaction
FU	functional unit
GHG	greenhouse gas
GW	global warming
LCA	life cycle assessment
LHV	lower heating value
LPC	levelized product costs
NG	natural gas
PHS	pumped hydro storage
PSA	pressure swing adsorption
rWGS	reverse water-gas-shift
SMR	steam-methane-reforming
SNG	synthetic natural gas
VRB	vanadium redox flow battery

Chemicals

CH_4	methane
CH_3OH	methanol
CO	carbon monoxide
CO_2	carbon dioxide
H_2	hydrogen
H_2CO	formaldehyde
$HCOOH$	formic acid
$HCOOCH_3$	methyl formate
H_2O	water
N_2O	nitrous oxide
NEt_3	triethylamine
$NHex_3$	trihexylamine
NR_3	tertiary amine

Greek symbols

η	efficiency

Kurzfassung

In den letzten Jahren hat das Interesse an der Umwandlung von CO_2 zu Basischemikalien mit einem Kohlenstoffatom (C1-Chemikalien) wie z.B. Methan und Methanol stark zugenommen. Die Motivation für die Nutzung von CO_2 ist die Reduktion von Treibhausgasemissionen und des fossilen Rohstoffverbrauchs. Diese Reduktionen sind jedoch nicht garantiert, da alle C1-Chemikalien neben dem im Überfluss vorhandenen CO_2 auch Wasserstoff benötigen. Daher ist das Ziel dieser Arbeit die ökologische Bewertung von CO_2-basierten C1-Chemikalien (Methan, Methanol, Kohlenstoffmonoxide und Ameisensäure) auf Basis einer Ökobilanz (engl. life cycle assessment, LCA). Die Bewertung erfolgt aus einer system-weiten Perspektive, das heißt, für limitierte Rohstoffe (z.B. erneuerbarer Strom) wird auch berücksichtigt, wie der limitierte Rohstoff sonst in anderen Prozessen genutzt worden wäre.

Zunächst werden die CO_2-basierten Prozesse mit den derzeitigen fossilen Prozessen für C1-Chemikalien verglichen. Ameisensäure bietet das höchste Potential zur Einsparung von Treibhausgasemissionen und fossilen Rohstoffen. Danach folgen Kohlenstoffmonoxid, Methanol und Methan. Für Ameisensäure ist sogar die derzeitige fossile Bereitstellung von Wasserstoff ausreichend, um Treibhausgasemissionen und fossile Rohstoffe einzusparen. Alle anderen Prozesse benötigen Wasserstoff aus einer Elektrolyse mit erneuerbarem Strom.

Im Folgenden wird die Bereitstellung von Wasserstoff durch eine Elektrolyse genauer untersucht. Die CO_2-basierte Herstellung von Kohlenstoffmonoxid und Methan benötigt etwa 60 % bzw. 88 % erneuerbaren Strom (in 2020 in der EU-27) um Treibhausgasemissionen gegenüber den fossilen Prozessen zu sparen.

Wenn 100 % erneuerbarer Strom genutzt wird, reduzieren alle CO_2-basierten C1-Chemikalien Treibhausgasemissionen und den fossilen Rohstoffverbrauch im Vergleich zu fossilen Prozessen. Zur Bewertung dieser Reduktionen werden in dieser Arbeit ebenfalls alternative Möglichkeiten zur Integration von erneuerbarem Strom (Power-to-X) ökologisch bewertet wie z.B. Stromspeicher, Batteriefahrzeuge und Wärmepumpen. Die höchsten Reduktionen pro genutztem Strom werden durch die Bereitstellung von Wärme erzielt, gefolgt von Batteriefahrzeugen und klassischen Stromspeichern. Erst danach folgen die CO_2-basierten C1-Chemikalien.

Da erneuerbarer Strom effizienter außerhalb der chemischen Industrie genutzt werden kann, werden ebenfalls biomassebasiertes Methan und Methanol untersucht. Biomasse erzielt die höchsten Einsparungen, wenn es Kohlekraftwerke ersetzt, dann folgt die Herstellung von Methanol. Für Methanol und Methan kann der Ertrag von Biomasseprozessen durch die zusätzliche Nutzung von Wasserstoff deutlich erhöht werden.

Abstract

In recent years, the conversion of CO_2 to basic chemicals with one carbon atom (C1-chemicals) such as methane and methanol has gained increasing interest. The major motivation for the utilization of CO_2 is the reduction of global warming and fossil depletion impacts. However, these reductions are not guaranteed because all C1-chemicals require hydrogen besides the abundantly available CO_2. Thus, the goal of this thesis is the life cycle assessment of CO_2-based C1-chemicals (methane, methanol, carbon monoxide and formic acid). The assessment is based on a system-wide perspective, which means that for limited resources such as renewable electricity also the utilization of the limited resources is in other processes is considered.

First of all, the CO_2-based processes are compared to fossil-based processes for C1-chemicals. Formic acid has the highest potential to reduce global warming and fossil depletion impacts followed by carbon monoxide, methanol and methane. Even if hydrogen is supplied by fossil-based steam reforming, formic acid reduces global warming and fossil depletion impacts. All other CO_2-based C1-chemicals require hydrogen from electrolysis using renewable electricity.

In the following, the supply of hydrogen by electrolysis is analyzed in more detail. The CO_2-based processes for carbon monoxide and methane required about 60 % and 88 % renewable electricity (in 2020 in the EU-27) to reduce global warming impacts compared to the fossil-based processes.

If 100 % renewable electricity is used, all CO_2-based C1-chemicals reduce global warming and fossil depletion impacts compared to the fossil-based processes. For the assessment of these reductions, also alternative utilization options for renewable electricity (Power-to-X) are analyzed such as electricity storage systems, battery electric vehicles and heat pumps. The highest reductions per electricity used are achieved for heat pumps followed by battery electric vehicles and electricity storage systems. Then, the CO_2-based C1-chemicals follow.

Since renewable electricity is used more efficiently outside the chemical industry, also biomass-based methane and methanol are analyzed. The utilization of biomass achieves the highest reductions if coal-fired power plants are substituted followed by the production of methanol. For methanol and methane, the yield per biomass can be increased if additional hydrogen is used.

Chapter 1

Introduction

Since the industrial revolution in the 18th century, fossil resources such as coal, petroleum and natural gas are the fuel for economic growth (Olah, 2005; Shafiee and Topal, 2009). In the beginning of the industrial revolution, coal was the predominant fossil resource, which, for example, was used as fuel for steam engines (Olah, 2005). Since the second half of the 19th century, petroleum and natural gas have additionally emerged as important fossil-based energy carriers (Olah, 2005). Today, fossil resources are used as fuel for electricity and heat generation and for transportation (Economides and Wood, 2009). Additionally, fossil resources are carbon carriers and thus are used as raw materials for petrochemicals (Economides and Wood, 2009).

The utilization of fossil resources has grown continuously since the industrial revolution and has reached about 490 EJ in 2013 (IEA, 2016). This increasing utilization of fossil fuels causes 2 major environmental problems: (I) depletion of fossil resources and (II) increasing carbon dioxide (CO_2) content in the atmosphere (Olah, 2005). The CO_2 produced during combustion of fossil fuels is considered as main driver of global warming (Pachauri et al., 2014).

To fight these environmental problems, a shift is required from fossil resources to renewable resources such as solar, hydro, biomass, wind, ocean and geothermal energy (Moriarty and Honnery, 2016; REN21, 2015). Currently, the integration of renewable energy sources is most advanced for electricity generation. Almost 60 % of the worldwide power generation capacity installed in 2014 was based on renewable energy sources (REN21, 2015). For the current global heat demand, the share of renewable energy sources is about 25 % (REN21, 2015). Here, the majority of renewable energy is supplied by traditional biomass. Heat from renewable resources can also be supplied by solar and geothermal energy. For transportation fuels, the substitution of fossil resources is more challenging because renewable resources such as solar, hydro, wind are no suitable energy carriers. Biomass can be considered as energy carrier but compared to fossil resources, biomass has a low energy density. Today, renewable resources account for about 3.5 % of the energy demand in the transportation sector. The majority of this renewable energy is supplied by liquid fuels derived from biomass. A further strongly emerging technology that enables the utilization of renewable electricity is battery electric vehicle. To substitute fossil resources as raw materials for petrochemicals, renewable carbon carriers are required. The only renewable resource which contains carbon is biomass. However, the amount of sustainably useable biomass is very limited and the majority of biomass is currently used for electricity generation (Vennestrøm et al., 2011).

Thus, alternative renewable carbon carriers are required. Besides biomass, CO_2 is a further renewable carbon source. In contrast to biomass, CO_2 is abundantly available (von der Assen et al., 2016). Recently, the utilization of CO_2 as carbon source gained increasing attention (Peters et al., 2011; Aresta et al., 2013; Perathoner and Centi, 2014). In particular, the CO_2-based production of formic acid (Moret et al., 2014; Jens et al., 2016), carbon monoxide (Jafarbegloo et al., 2015), methanol (Jadhav et al., 2014) and methane (Rönsch et al., 2016) (C1-chemicals) has gained much interest. For these C1-chemicals, first pilot plants are already available (Kreimeyer, 2013; Jafarbegloo et al., 2015; Jadhav et al., 2014; Rönsch et al., 2016). These CO_2-based C1-chemicals are not only carbon carriers, they are also energy carriers and can be used over the same variety of applications as fossil resources. In particular, CO_2-based methane and methanol also offer the option to substitute fossil-based transportation fuels (Olah, 2005).

However, while CO_2 utilization introduces an essentially renewable carbon source, an environmental benefit cannot be taken for granted for every CO_2 utilization process. In particular, the supply of hydrogen, which is required as co-reactant, is associated with environmental impacts. A suitable method to evaluate the environmental impacts is life cycle assessment (LCA). Recent LCA studies for methanol, methane and CO_2-based fuels already show that a reduction of global warming and fossil depletion impacts compared to corresponding conventional processes can be achieved if renewable electricity is used for hydrogen production (von der Assen et al., 2013; van der Giesen et al., 2014; Reiter and Lindorfer, 2015).

However, renewable electricity is currently very limited and almost exclusively used to replace fossil-based power plants. Until 2050, the share of renewable energy sources in the EU-27 electricity supply is expected to rise to about 50 %. Thus, even in 2050, the availability of renewable electricity will still be limited. In addition, renewable electricity is required by numerous applications proposed in the last years for heating, transportation and chemicals. Thus, it is not only important to achieve lower impacts than fossil-based processes from using valuable renewable electricity. In addition, the

CO_2-based processes must be ranked according to their environmental benefit of using renewable electricity. Thus, the major goal of this thesis is to present an approach that allows to compare the environmental benefit of CO_2-based C1-chemicals with alternative utilization options for renewable electricity.

1.1 Structure of this thesis

Chapter 2 introduces the considered C1-chemicals and the corresponding CO_2 conversion reactions. For the environmental assessment of CO_2-based C1-chemicals, life cycle assessment (LCA) is introduced as suitable methodology. Subsequently, a review of LCA studies for CO_2-based production of C1-chemicals shows that challenges remain for a sound environmental assessment.

In Chapter 3, the contribution of this thesis – the system-wide perspective for LCA – is presented. For this purpose, the system-wide interactions of the CO_2-based processes for C1-chemicals with other processes are presented first. Then, it is explained how theses interactions are taken into account in this thesis.

Chapter 4 forms the basis of this thesis (Figure 1.1). Herein, the analyzed process concepts for formic acid, carbon monoxide, methanol and methane are presented for both CO_2-based (CO_2-to-Chemicals) and fossil-based processes. Furthermore, the maximum reduction of environmental impacts is introduced for CO_2-based production of C1-chemicals. The maximum reduction of environmental impacts is the difference of environmental impacts for fossil-based processes and CO_2-based processes without considering the environmental impact of hydrogen supply ("free" hydrogen). Based on the maximum reduction of environmental impacts, (i) supply processes for hydrogen and CO_2 that lead to environmentally beneficial CO_2-based processes can be identified and (ii) the CO_2-based processes can be compared according to their environmental benefit per kg hydrogen used. In Chapter 4, the CO_2-based C1-chemicals are considered exclusively as raw material for the chemical industry. The results show that the CO_2-based production of carbon monoxide, methanol and methane require hydrogen supply by renewable electricity to achieve lower environmental impacts than the fossil-based processes.

In Chapter 5, the hydrogen supply by water electrolysis is considered in detail including steady-state and part-load operation, and hydrogen storage. For this case study, the CO_2-based production of carbon monoxide and methane is considered (Power-to-Gas). Carbon monoxide and methane represent a best case (carbon monoxide) and a worst case (methane) for CO_2-based processes that require mandatorily

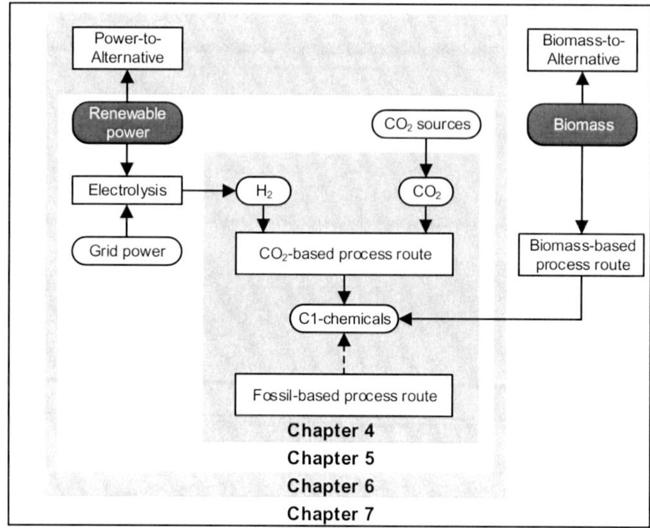

Figure 1.1: Expanding scope for LCA of CO_2-based processes for C1-chemicals as taken in the chapters of this thesis.

renewable electricity to be environmentally beneficial. In this case study, CO_2 is supplied by a coal-fired power plant with CO_2 capture.

In Chapter 6, the scope of the environmental assessment is expanded. Here, the utilization of renewable electricity for the CO_2-based production of carbon monoxide, methanol and methane is compared to alternative utilization options for renewable electricity (Power-to-X systems). Furthermore, the CO_2-based C1-chemicals are considered as energy carriers for electricity generation (Chemical-to-Power) as well as for transportation purposes (Chemical-to-Mobility).

In Chapter 7, the scope of the environmental assessment is expanded to include biomass-based production of C1-chemicals. An optimization-based approach is presented to identify environmentally most beneficial processes for the production of C1-chemicals depending on the alternative utilization options of the renewable inputs biomass and renewable electricity. The approach is presented by a case study for methanol and methane production.

In Chapter 8, the thesis is summarized and conclusions are drawn. Furthermore, topics for future research are identified.

Chapter 2

Current status of the environmental assessment of CO_2-based C1-chemicals

This chapter provides a general overview of CO_2 conversion reactions to C1-chemicals and their environmental assessment. In Section 2.1, the term C1-chemicals is defined and an overview of CO_2 conversion reactions is given. In Section 2.2, life cycle assessment (LCA) is introduced as suitable method to determine the environmental consequences due to the introduction of CO_2-based processes for C1-chemicals. Section 2.3 reviews LCA studies of CO_2 conversion to C1-chemicals and presents remaining challenges.

Parts of this chapter are reproduced with permission from The Royal Society of Chemistry from:

> Sternberg, A., Jens, M.C. and Bardow, A. (2017). Life Cycle Assessment of CO_2-based C1-chemicals, *Green Chemistry*, 19(9):2244-2259.

2.1 Introduction to CO_2 conversion to C1-chemicals

In this section, the term C1-chemicals is defined and the corresponding CO_2-based reactions are presented. Process concepts including the purification of products are presented in Section 4.2.

2.1.1 Definition and utilization of C1-chemicals

In this thesis, the term C1-chemicals denotes *all compounds with 1 carbon atom that can be produced exclusively from CO_2 and hydrogen.* According to this definition, the following C1-chemicals exist:

- carbon monoxide (CO),
- formic acid (HCOOH),
- formaldehyde (H_2CO),
- methanol (CH_3OH) and
- methane (CH_4).

In the following, the importance of C1-chemicals is highlighted by presenting their current and potential future utilization.

Carbon monoxide. For carbon monoxide, one can distinguish the utilization as pure component or as a constituent of syngas (mixture of primarily hydrogen and carbon monoxide). Pure carbon monoxide is used for the preparation of acetic acid, as reducing agent in blast furnaces, as reactant for phosgene production, and for the production of formic acid and methyl formate. Syngas is required for the production of a large variety of chemicals, e.g., methanol, hydrocarbons and linear aliphatic aldehydes (Ledon, 1986).

A potential future application for syngas is the Fischer-Tropsch reaction. Currently, the Fischer-Tropsch reaction is gaining much attention as option to produce synthetic liquid fuels (Kaiser et al., 2013; van der Giesen et al., 2014; Baltrusaitis and Luyben, 2015). The synthetic liquid fuels can be used to substitute petroleum-based liquid fuels.

Formic acid. Formic acid is currently used for a variety of applications including: silage and animal feed (27 %), leather and tanning (22 %), pharmaceuticals and food chemicals (14 %), textile (9 %), natural rubber (7 %), drilling fluids (4 %) and others (17 %) (Hietala et al., 2016).

Furthermore, formic acid is considered as potential future hydrogen carrier (Leitner, 1995; Supronowicz et al., 2015; Klankermayer et al., 2016). The benefit of formic acid is that according to the stoichiometric reaction no hydrogen is sacrificed for water production during the reaction.

Formaldehyde. Formaldehyde is a crucial intermediate for the synthesis of many other chemicals. The majority of formaldehyde is used for the production of resins (e.g., urea-formaldehyde, melamine-formaldehyde and phenol-formaldehyde); polyoxymethylene (POM), methylene diphenyl diisocyanate (MDI) and 1,4-butanediol (BD) (Reuss et al., 2000).

Methanol. About 70 % of methanol is used for synthesis of chemicals. The most important are formaldehyde, methyl tert-butyl ether, acetic acid, methyl methacrylate and dimethyl terephthalate. Furthermore, methanol is used as solvent, e.g., for the production of formic acid (Fiedler et al., 2000).

Currently, the energetic utilization of methanol is of minor importance. However, options for the utilization of methanol as energy carrier have already been shown by Asinger (1986). Methanol can be used in Otto engines, diesel engines and fuel cells.

Methane. Methane is the main constituent of natural gas. Natural gas is used for generation of electricity and heat, and as feedstock for chemicals. In particular, the current industrial production of all previously presented C1-chemicals is mainly based on natural gas (Hammer et al., 2000).

2.1.2 Overview of CO_2 conversion reactions to C1-chemicals

An overview of CO_2 conversion reactions to C1-chemicals is presented in Figure 2.1. First pilot plants for CO_2-based production are currently available for formic acid (Kreimeyer, 2013), carbon monoxide (Jafarbegloo et al., 2015), methanol (Jadhav et al., 2014) and methane (Rönsch et al., 2016). So far, a process concept for the catalytic one-step reaction from CO_2 to formaldehyde is not available. However, formaldehyde can be produced from CO_2 by a combination of CO_2-based methanol production and the conventional methanol to formaldehyde process (von der Assen et al., 2015).

For the fossil-based routes, all C1-chemicals need stoichiometrically 1 mol methane. Syngas and carbon monoxide can be produced in a one-step reaction from methane. Carbon monoxide is used as feedstock for formic acid production. Syngas is used as feedstock for methanol production. In turn, methanol is used to produce formaldehyde. Since each reaction step has conversion losses and requires additional process

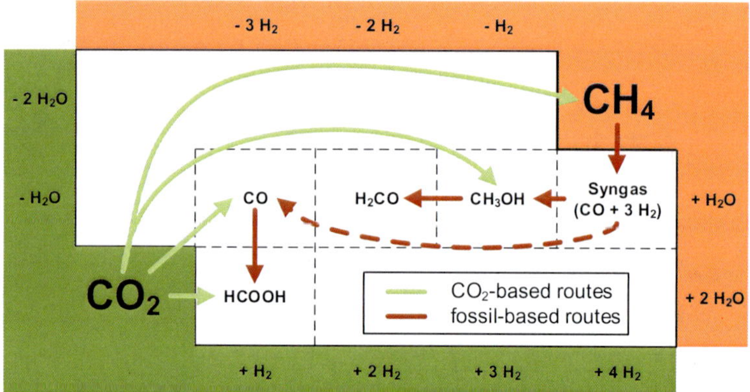

Figure 2.1: Overview of C1-chemicals and their CO_2-based and fossil-based routes. For CO_2-based routes, the bottom axis shows the amount of hydrogen required stoichiometrically and the left axis the amount of produced water. For fossil-based routes, the right axis shows the stoichiometrically required amount of water and the top axis the amount of produced hydrogen. Solid lines represent reactions and the dashed line purification processes.

energy, the overall energy demand per mole C1-chemical increases with the number of reaction steps. Here, the CO_2-based routes offer an advantage: The C1-chemicals can be produced in one step from CO_2. Thus, CO_2-based formic acid and methanol production seems most promising, because the number of reaction steps is decreased compared to fossil-based processes.

However, the stoichiometric reaction (Figure 2.1) should only be considered indicative for the environmental potential of CO_2-based processes for C1-chemicals. For a reliable assessment, more information than just the stoichiometric reaction is required: Consider the advantage that no hydrogen is lost in the CO_2-based production of formic acid. This advantage is also a disadvantage: The creation of water from hydrogen is thermodynamically favorable, and thus drives the reaction yield. For this purpose, the following comparative life cycle assessment is based on process simulations considering thermodynamic properties. The considered processes are presented in the next subsection.

Currently, the most mature CO_2 conversions to C1-chemicals are based on catalytic reactions of CO_2 with hydrogen, so-called hydrogenation reactions (Kondratenko et al., 2013; Ampelli et al., 2015).

CO_2 hydrogenation to formic acid. The equimolar conversion of CO_2 and hydrogen yields formic acid:

$$CO_2(g) + H_2(g) \rightleftharpoons HCOOH(l). \tag{2.1}$$

The reaction of gaseous CO_2 and hydrogen to formic acid is endergonic (Δg°_R = 32.7 kJ/mol), therefore, the equilibrium is on the left side of the reaction. The equilibrium can be shifted to the product side, if the reaction takes place in aqueous solution (Moret et al., 2014):

$$CO_2(aq) + H_2(aq) \rightleftharpoons HCOOH(aq). \tag{2.2}$$

An exergonic reaction can also be achieved by adding bases such as tertiary amines (NR_3) (Schaub and Paciello, 2011). Proposed tertiary amines are for example trihexylamine ($NHex_3$) (Schaub and Paciello, 2011) and triethylamine (NEt_3) (Anderson et al., 1986; Green et al., 1989).

$$CO_2(g) + H_2(g) + NR_3(aq) \rightleftharpoons HCO_2^-(aq) + NR_4^+(aq), \tag{2.3}$$

$$HCO_2^-(aq) + NR_4^+(aq) \rightleftharpoons HCOOH(aq) + NR_3(aq). \tag{2.4}$$

In this case, CO_2 and hydrogen first react with a tertiary amine to an amine formic acid adduct (NR_3–HCOOH). In a subsequent process step, the amine formic acid adduct is converted to formic acid and the tertiary amine. The CO_2-based production of formic acid with trihexylamine has already been tested in a pilot plant by BASF (Kreimeyer, 2013).

CO_2 hydrogenation to carbon monoxide. The reaction of 1 mole CO_2 and 1 mole hydrogen can also yield carbon monoxide. This so-called reverse water-gas-shift (rWGS) reaction produces water as by-product:

$$CO_2 + H_2 \rightleftharpoons CO + H_2O. \tag{2.5}$$

The production of carbon monoxide in the rWGS reaction is favored at high temperatures and low pressures (Kaiser et al., 2013). A pilot plant for the rWGS reaction is operated by Sunfire (Verdegaal et al., 2015).

Carbon monoxide can also be produced from CO_2 and methane by dry reforming of methane (DRM). The DRM reaction produces carbon monoxide and hydrogen with a molar ratio of 1:

$$CO_2 + CH_4 \rightleftharpoons 2H_2 + 2CO. \tag{2.6}$$

The DRM reaction is favored at high temperatures and low pressures. For this reaction, also first pilot plants are available (Jafarbegloo et al., 2015).

CO_2 hydrogenation to formaldehyde. The reaction of 1 mole CO_2 and 2 mole hydrogen yields formaldehyde:

$$CO_2 + 2H_2 \rightleftharpoons H_2CO + H_2O. \tag{2.7}$$

For CO_2-based production of formaldehyde, fundamental research is still required to achieve an industrial process. Thus, this reaction is not further considered for the environmental assessment of C1-chemicals. Currently, most CO_2 conversions to formaldehyde proposed are based on reduction of CO_2 on metal hydride complexes instead of using molecular hydrogen (Bontemps et al., 2014; Jiang et al., 2013; Pal et al., 2015; Rios et al., 2016).

CO_2 hydrogenation to methanol. Methanol can be produced from CO_2 and hydrogen according to the following reaction:

$$CO_2 + 3H_2 \rightleftharpoons CH_3OH + H_2O. \tag{2.8}$$

The methanol production is favored at low temperatures and high pressures. The CO_2-based production of methanol has been tested in several pilot plants (Jadhav et al., 2014). Furthermore, a commercial plant exists in Iceland (Jadhav et al., 2014).

CO_2 hydrogenation to methane. 1 mole CO_2 can react with 4 mole hydrogen to methane according to the Sabatier reaction:

$$CO_2 + 4H_2 \rightleftharpoons CH_4 + 2H_2O. \tag{2.9}$$

The Sabatier reaction is favored at low temperatures. A list of pilot plants is presented in Rönsch et al. (2016).

2.2 Introduction to life cycle assessment (LCA)

The major motivation for utilization of CO_2 as feedstock in the chemical industry is the substitution of fossil feedstocks such as oil and natural gas (Centi et al., 2013; Kondratenko et al., 2013; Aresta et al., 2013). Furthermore, CO_2 utilization can avoid CO_2 emissions, for example, if CO_2 replaces other feedstocks with higher emissions (von der Assen and Bardow, 2014). CO_2 emissions are the primary cause of anthropogenic global warming. A suitable method to benchmark fossil depletion and global warming impacts of CO_2-based processes with fossil-based processes is life cycle assessment (LCA) (von der Assen et al., 2014).

LCA is a systematic method for the environmental assessment of processes and products (ISO 14044, 2006; ISO 14040, 2009). For the environmental assessment, interactions of products or services with the environment are analyzed (Figure 2.2). The interactions with the environment are called elementary flows and can be divided into inputs (use of resources) and outputs (emissions). The interactions can be analyzed along the entire life cycle of a product, i.e., from raw material extraction to waste treatment (cradle-to-grave), or for a segment of the life cycle, e.g., from raw material extraction to product supply (cradle-to-gate). For the interpretation of the interactions with the environment, several environmental impact categories are available such as global warming, resource depletion, eutrophication, acidification and toxicity. Each environmental impact category weights the elementary flows that affect the impact category. For example, the environmental impact category *global warming* weights all elementary flows that contribute to global warming such as CO_2, CH_4 and N_2O emissions.

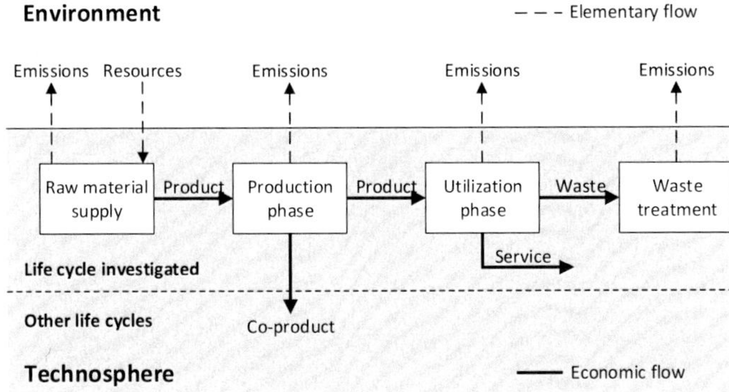

Figure 2.2: System boundaries for LCA

A comprehensive description of LCA can be found in ISO 14044 (2006) and ISO 14040 (2009) on which the following discussion is based. In the following, only aspects for LCA that are crucial for the comprehension of this thesis are explained in more detail.

Application. The environmental impacts determined by an LCA can be used (I) to compare different products or processes that supply the same function (so-called comparative LCA) or (II) for accounting purposes, e.g., environmental product declarations and carbon footprints. However, ultimately, results of LCA studies with accounting purposes are also used to compare different processes or products, e.g., by customers (Weidema, 2000). In this thesis, LCA is used to *compare* CO_2-based processes for C1-chemicals with competing processes, e.g., fossil-based processes.

Functional unit. To ensure that processes are compared on an equal basis, a reference unit must be defined. This reference unit is called functional unit in LCA. The functional unit usually contains the product or service under study.

Multi-functionality. In LCA, processes that supply more than 1 function (product or service) are denoted multi-functional. The problem of multi-functionality in LCA is that the environmental impacts of a multi-functional process, which supplies for example 2 functions are shared by both functions. If the life cycle investigated only contains 1 of these functions, not the total environmental impact of the multi-functional process should be included.

In LCA, 3 options exist to fix the multi-functionality problem: (i) allocation, (ii) system expansion and (iii) avoided burden.

(i) The allocation method transforms multi-functional processes into *quasi*-mono-functional processes. In this case, the environmental impacts of the multi-functional process are allocated to the *quasi*-mono-functional processes based on allocation factors. The allocation factors can be derived by the relationship (e.g., mass, economic value or exergy) between the functions of the process. Depending on the chosen relationship, the allocation factors can vary.

(ii) The system expansion method expands the functional unit. In other words, the function of the multi-functional process that is not used in the life cycle investigated is also included. If system expansion is used for a comparative LCA study, the functional unit of all compared life cycles must contain the same functions. If functions are missing, mono-functional processes must be added that supply the missing function.

(iii) The avoided burden method subtracts the function of the multi-functional process that is not used in the life cycle investigated by a mono-functional process for this function. In this case, the functional unit is not affected.

System boundaries. The system boundaries describe which processes are include in an LCA. Typically, processes are included in the system boundaries that are connected with the product under study by energy or matter flows. If multi-functional processes occur in the system boundaries, the co-product that is not within the scope of the study can be excluded from the system boundaries by allocation or avoided burden, or the utilization of the co-product can be integrated by system expansion.

The system boundaries also describe the considered life-cycle stages of a product. For example, if the goal of an LCA is to compare different processes for equal products, the identical utilization phase and end-of-life stage can be excluded (cradle-to-gate). The consideration of the entire life-cycle is for example required if equal services are compared, e.g., transportation of 1 l milk. In this case, the end-of-life treatment can be different (e.g., for bottles and cartons) and thus should be considered (cradle-to-grave).

2.2.1 Attributional LCA (ALCA) vs consequential LCA (CLCA)

In the literature, LCA is often classified into 2 approaches: attributional and consequential LCA. Both approaches are briefly presented in the following.

Definition ALCA. Finnveden et al. (2009) defined ALCA "by its focus on describing the environmentally relevant physical flows to and from a life cycle and its subsystems."

Thus, in ALCA, static processes with average historical data are used and allocation is typically applied to fix multi-functionality problems (Thomassen et al., 2008). Consequently, an ALCA study cannot inherently capture potential consequences in other life cycles. However, a consequence in another life cycle can be considered through the comparison of different life cycles (Figure 2.3). Typically, a new process (i.e., Life cycle A in Figure 2.3) for a product is compared to a conventional existing process (i.e., Life cycle B in Figure 2.3).

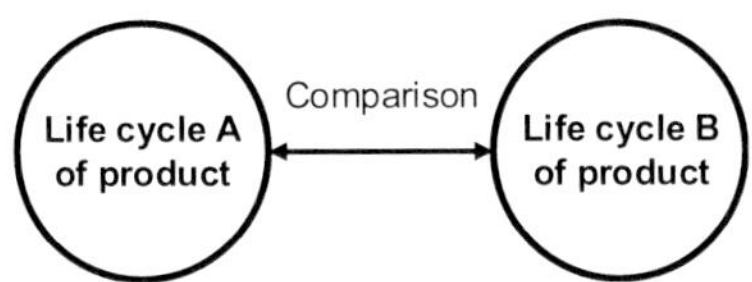

Figure 2.3: Comparison of processes with ALCA

Definition CLCA. According to Finnveden et al. (2009), "CLCA is defined by its aim to describe how environmentally relevant physical flows will change in response to possible decisions."

Thus, in CLCA, affected processes by change in demand are considered in the life-cycle investigated but also in other affected life-cycles (Figure 2.4). The life cycle investigated can interact with other life cycles due to technological effects (flows of energy and matter), market effects and behavioral effects (Kätelhön et al., 2015).

An important feature of CLCA is the identification of affected processes based on models of the economic system (Rajagopal, 2017). However, identifying all affected processes by a decision is almost impossible. Most current CLCA studies only identify one affected process (Zamagni et al., 2012). Thus, Zamagni et al. (2012) stated that "CLCA is still far from a proper systematization".

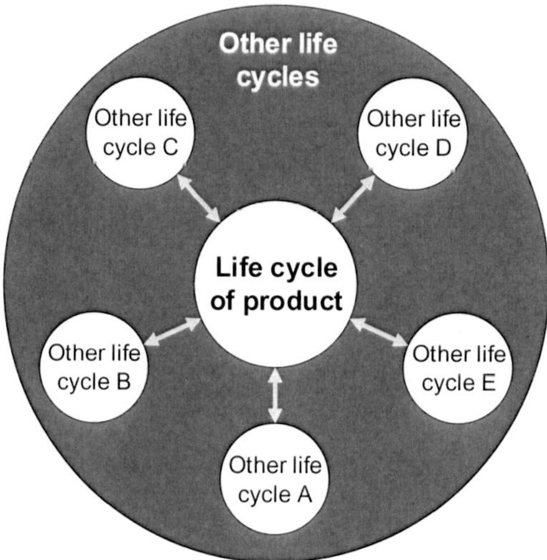

Figure 2.4: Comparison of processes with CLCA

Combination of ALCA and CLCA. In practice, both approaches are often combined. For example, in ALCA, also often system expansion or avoided burden is used to treat multi-functionality problems (Thomassen et al., 2008). Through the utilization of system expansion or avoided burden, interactions with further life cycles can be considered in ALCA. If a CLCA study uses data sets from an LCA database usually allocation is included in some background processes. In this case, the interaction of the life cycle investigated with background processes is not considered. The following section reviews which interactions of CO_2-based processes with other life cycles are taken into account in present LCA studies.

2.3 Review of LCA studies for CO_2 conversion to C1-chemicals

So far, LCA studies for CO_2 conversion are available for the following C1-chemicals: methane, methanol and carbon monoxide (Table 2.1). For carbon monoxide, only the utilization as precursor for the production of liquid hydrocarbon fuels is considered. An LCA study is so far missing that investigates the utilization of CO_2-based carbon monoxide in the chemical industry. For formic acid, there is no LCA study at all.

Table 2.1: Present LCA studies for CO_2 conversion to C1-chemicals

Product	Reference
methane	Reiter and Lindorfer (2015)
methanol	Aresta et al. (2002); Dumont et al. (2012) and von der Assen et al. (2013)
carbon monoxide	van der Giesen et al. (2014)

In the following, current LCA studies for CO_2-based C1-chemicals are reviewed. The focus of the review is on interactions with other life cycles in the LCA studies rather than on the LCA results themselves. For this purpose, in particular, hydrogen and CO_2 supply are reviewed for the CO_2-based production of C1-chemicals. Furthermore, the considered benchmark processes are analyzed.

CO_2-based processes

In all published LCA studies for C1-chemicals, the CO_2-based processes consist of hydrogen supply, CO_2 supply and a CO_2 conversion process. In the CO_2 conversion process, CO_2 and hydrogen are used to produce the C1-chemical.

CO_2 conversion processes. In current LCA studies, all considered CO_2 conversion processes are mono-functional. For example, the utilization of heat from exothermic CO_2-based methane production is not considered. Thus, no interactions with other life cycles are considered. The only inputs are CO_2 and hydrogen. Reiter and Lindorfer (2015) and van der Giesen et al. (2014) considered the inputs and outputs of the CO_2 conversion process based on stoichiometric data. Aresta et al. (2002) used data from a pilot plant.

CO_2 supply. The supply of CO_2 can be multi-functional (e.g., CO_2 capture from flue gas) or mono-functional (e.g., air capture).

I) *CO_2 capture from flue gas* is considered in all published LCA studies for C1-chemicals. However, the treatment of the multi-functionality problem is different in all LCA case studies.

Reiter and Lindorfer (2015) consider CO_2 capture by amine scrubbing from flue gas. The process where CO_2 is formed is not included in the system boundaries. Only the additional energy demand due to the CO_2 capture process is allocated to the CO_2 supply. However, the reduction of CO_2 emissions is not allocated to the CO_2 during the supply (life-cycle) stage. Instead, CO_2 emissions are neglected in the end-of-life stage of the C1-chemical if CO_2 has a biogenic origin or if CO_2 has a fossil origin and would otherwise be emitted. If fossil-based CO_2 would alternatively be stored (CCS), the CO_2 emissions in the end-of-life stage are considered. For biogenic CO_2, such a distinction is missing. However, such a distinction should also be made for biogenic CO_2: CO_2 emissions are also not avoided due to CO_2 utilization if the utilization of biogenic CO_2 is compared to storage of biogenic CO_2. This problem can be avoided by explicitly counting CO_2 at each life cycle stage as suggested by Rabl et al. (2007) for biomass-based processes.

Van der Giesen et al. (2014) considered CO_2 capture from biomass-fired and fossil-fired power plants. These authors include the power plants with CO_2 capture in the system boundaries. Both products electricity and captured CO_2 are considered as inputs for the CO_2-based process. The electricity is used to satisfy a minor amount of electricity demand for electrolysis. Through the internal utilization of the by-product electricity, no other life cycle is affected. In this case, the utilization of CO_2 does not avoid CO_2 emissions if a fossil-fired power plant is used. Thus, the CO_2-based process only reduces CO_2 emissions compared to the fossil-based processes if a biomass-based power plant is used.

Von der Assen et al. (2016) compared the impacts for global warming and fossil depletion of various CO_2 capture processes, e.g, power plants and industrial processes. The utilization of the captured CO_2 was not within the scope of the study. All processes with CO_2 capture are compared to the corresponding process without CO_2 capture (status-quo). Here, the environmental impact of captured CO_2 is based on the additional environmental burden of CO_2 capture (e.g., additional energy demand) and the additional environmental benefit of CO_2 capture (CO_2 is not emitted). All CO_2 capture processes lead to reductions of CO_2 emissions.

All considered LCA studies have in common that no allocation is used for multi-functional CO_2 supply processes. Rather, system expansion or avoided burden is used to include the interaction with the life cycle of CO_2 supply in the LCA study.

II) *Air capture* is a mono-functional process. Thus, the environmental impact of CO_2 supply results from the energy demand of the process and the CO_2 uptake from the air. Air capture for CO_2 supply is considered by von der Assen et al. (2016) and van der Giesen et al. (2014). Interestingly, due to different treatment of multi-functionality for CO_2 capture from flue gases, von der Assen et al. (2016) conclude that CO_2 supply by air capture is least efficient in global warming terms, while van der Giesen et al. (2014) conclude CO_2 supply by air capture is the only sustainable CO_2 source.

The different treatments of CO_2 supply in current LCA studies show that clarification is still required for the multi-functionality problem.

Hydrogen supply. In current LCA studies, hydrogen supply is considered by steam-methane-reforming with water-gas-shift (Aresta et al., 2002) and water electrolysis (Aresta et al., 2002; Reiter and Lindorfer, 2015; van der Giesen et al., 2014).

I) *Hydrogen supply by steam-methane-reforming* is a mono-functional process. Using hydrogen supply by steam-methane-reforming, the environmental impacts of all CO_2-based processes that are analyzed until now (methane, methanol and carbon monoxide) are higher than for the corresponding fossil-based production (Reiter and Lindorfer, 2015; Aresta et al., 2002; van der Giesen et al., 2014).

II) *Hydrogen supply by electrolysis* is a multi-functional process. Besides hydrogen, water electrolysis also co-produces oxygen. However, in present LCA studies, the co-production of oxygen is not considered. Thus, water electrolysis is also treated as mono-functional process and no interactions are considered with the life cycle of oxygen supply.

For electricity supply to electrolysis, the following types of electricity are considered: grid mix and renewable electricity. In all available LCA studies, a reduction of environmental impacts for CO_2-based processes compared to fossil-based processes is only achieved if renewable electricity is employed. Reiter and Lindorfer (2015) and van der Giesen et al. (2014) mention that renewable electricity might be used with higher environmental benefits, for example, by replacing fossil-fired power plants. However, a systematic analysis for the interaction of CO_2-based processes with other life cycles using renewable electricity is so far missing. For example, van der Giesen et al. (2014) stated that "to assess if this [using CO_2] is a preferable route to store energy, additional research is needed in which the production and use of fuels from CO_2 needs to be compared with battery and hydrogen storage and possibly other storage options, such as water reservoirs."

Benchmark processes

In all present LCA studies, the CO_2-based processes are compared to fossil-based processes. All considered fossil-based processes are mono-functional. In the following, the benchmark processes chosen in the literature are reviewed for each C1-chemical.

Methane. For the CO_2-based production of methane, Reiter and Lindorfer (2015) consider natural gas on a lower heating value (LHV) basis as benchmark process. Reiter and Lindorfer (2015) also consider the utilization of CO_2-based methane as automotive fuel. In this case, the CO_2-based methane competes with diesel and gasoline.

Methanol. For the CO_2-based production of methanol, all LCA studies (Aresta et al., 2002; von der Assen et al., 2013; Dumont et al., 2012) consider the utilization as chemical and benchmark the CO_2-based process with natural-gas-based processes.

Carbon monoxide. van der Giesen et al. (2014) consider CO_2-based carbon monoxide as intermediate for the production of fuel. In this case, the CO_2-based fuel competes with diesel and gasoline. For the utilization of carbon monoxide in the chemical industry, no benchmark is considered.

While some authors consider biomass as CO_2 source (van der Giesen et al., 2014; Reiter and Lindorfer, 2015), biomass-based conversion processes are so far not considered as benchmark for the CO_2-based processes.

Main gaps in the literature

Based on the previous review, 4 main gaps are identified:

(1) case studies are not available for all C1-chemicals (HCOOH and CO),

(2) inconsistent treatment of CO_2 supply processes,

(3) alternative utilization of renewable electricity is not considered and

(4) comparison of CO_2-based processes to biomass-based processes is missing.

The present thesis aims at filling these gaps.

Chapter 3

Introducing a system-wide perspective for LCA

In current LCA studies, interactions of CO_2-based processes with other life cycles are considered for CO_2 supply and for fossil-based production of C1-chemicals. However, further important interactions of CO_2-based C1-chemicals with other life cycles are so far not considered, in particular, the competition for renewable inputs.

In this thesis, all interactions mentioned in the previous chapter are analyzed from a consequential perspective for the CO_2-based C1-chemicals. Since no economic modeling is used for quantification of affected processes, the term *system-wide perspective* is used rather than consequential LCA.

In Section 3.1, an overview is presented of all interactions considered for CO_2-based processes with other life cycles. Furthermore, the reason for not using economic modeling is presented. In Sections 3.2–3.4, the considered interactions are explained and it is presented how the interactions are considered. In Section 3.5, the maximum environmental impact reduction is introduced for CO_2-based C1-chemicals.

3.1 Interactions of CO_2-based processes with other life cycles

In this thesis, the CO_2-based processes for C1-chemicals are analyzed from a system-wide perspective. For this purpose, environmental impacts are considered in the life cycle of the CO_2-based processes **and** in other life cycles that interact with the CO_2-based processes (Figure 3.1). The following interactions of CO_2-based processes with other life cycles are investigated in this thesis:

- Competing processes for C1-chemicals such as fossil-based and biomass-based processes,
- Multi-functional processes such as CO_2 supply from power plants and
- Processes using renewable inputs such as renewable electricity, which is used for hydrogen supply in CO_2-based processes.

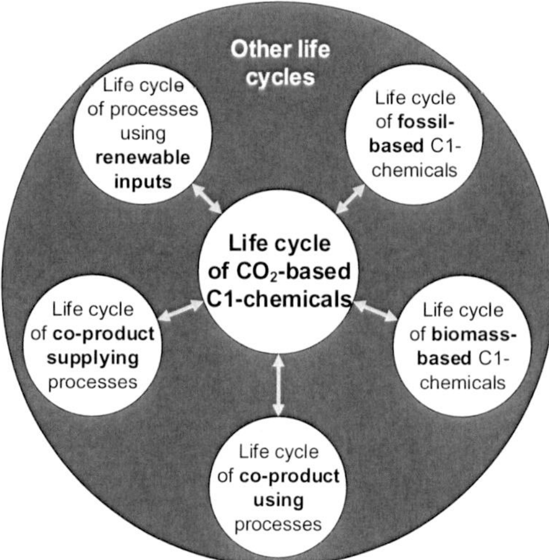

Figure 3.1: System-wide interactions of CO_2-based processes for C1-chemicals with other life cycles as considered in this thesis

In contrast to common applications of CLCA, no market-models are considered in this thesis for the quantification of consequences in each affected process. Instead, the full range of possible supply processes is considered for key processes (CO_2 supply, hydrogen supply and competing processes for C1-chemicals).

The benefits of considering a range of potentially affected supply processes instead of a particular supply process determined by economic modeling are

- taking into account the existing high uncertainty in economics of emerging technologies and
- considering the environmentally best option. This information is important for decision makers to set incentives for environmentally promising but expensive technologies.

In the following sections, it is explained how the interactions of CO_2-based processes with other life cycles are taken into account.

3.2 Competing processes for CO_2-based C1-chemicals

The introduction of CO_2-based processes for C1-chemicals causes an additional supply of C1-chemicals. The additional supply can either lead to (i) an increased utilization of the C1-chemical (increased demand) or to (ii) a reduced supply from a competing process (substitution) if the overall market demand stays constant. In the latter case, the CO_2-based processes cause consequences in the life cycle of the competing processes.

In this thesis, a constant market demand is considered for all investigated products. According to Kätelhön et al. (2015), the assumption of a constant market demand allows to consider the main effects of introducing a new technology while significantly reducing the amount of data required. Thus, the CO_2-based processes always substitute a competing process.

Fossil-based process. In this thesis, primarily the utilization of C1-chemicals in the chemical industry is analyzed. In this case, the CO_2-based processes substitute fossil-based processes for the production of C1-chemicals. In Chapter 6, additionally the utilization of C1-chemicals as energy carriers for electricity generation and for transportation purposes is considered. In this case, CO_2-based processes substitute fossil-based processes for the supply of energy carriers.

Biomass-based process In Chapter 7, the substitution of fossil-based processes by CO_2-based processes is compared to the substitution of fossil-based process by biomass-based processes.

3.3 Multi-functionality

The introduction of CO_2-based processes causes consequences in other life cycles if the CO_2-based processes (or the substituted fossil-based processes) use a subset of co-products from a multi-functional process. In this thesis, the following multi-functional supply processes are considered:

- hydrogen supply by water electrolysis (co-product: oxygen),
- hydrogen supply by chemical processes such as chloralkali process (co-product: e.g. chlorine) and
- CO_2 capture from flue gases (co-product depends on process, e.g. electricity, cement or chemical products).

Additionally, the CO_2-based processes or substituted fossil-based processes can also produce co-products that are used in other life cycles. In this thesis, the following co-products are considered:

- heat from CO_2-based processes (with exothermic reactions) and
- hydrogen from production of fossil-based carbon monoxide and formic acid.

Fixing multi-functionality (system expansion vs avoided burden)

In this thesis, the 2 approaches systems expansion and avoided burden are used to fix multi-functionality problems. The choice between system expansion and avoided burden does not affect the results of comparative LCA studies (Klöpffer and Grahl, 2009). Thus, both approaches can be used, and even be combined. The benefit of system expansion compared to avoided burden is a transparent depiction of the considered processes in the functional unit. However, if the functional unit includes too many processes, the interpretation of LCA results becomes difficult. In this case, avoided burden can be used to reduce the size of the functional unit. For example, if multi-functionality occurs in supply processes, the functional unit would include all considered co-products from different supply processes (e.g., CO_2 capture from sources with different co-products such as electricity, steal and iron, paper and pulp, and cement (von der Assen et al., 2016)).

In this thesis, system expansion is used for main products as suggested by Jung et al. (2013). For all by-products, avoided burden is used to treat multi-functionality.

The main challenge for the application of system expansion and avoided burden is to identify the affected processes in other life cycles. Recommendations for the identification of affected processes are presented in Ekvall and Weidema (2004). These recommendations are used in this thesis. According to Ekvall and Weidema (2004), the affected processes in other life cycles are different for multi-functional processes with combined production and joint production. Figure 3.2 gives an overview of multi-functional processes that are presented in more detail in the following.

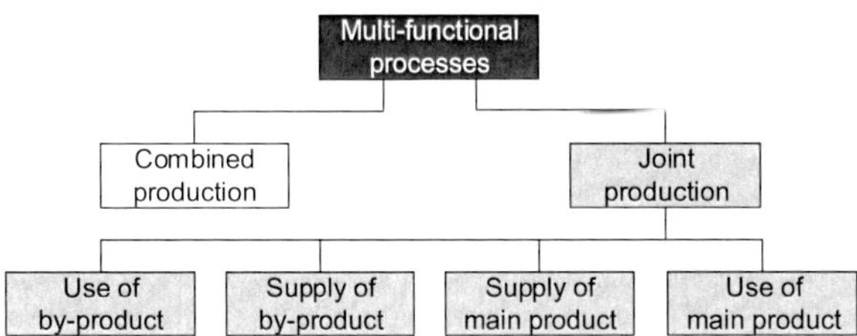

Figure 3.2: Overview of multi-functional processes considered in this thesis

Combined production

For the combined production of co-products, the ratio between co-products is variable. Thus, the supply of the required product for the CO_2-based process can be increased without affecting the supply of the other product (Figure 3.3). For combined production, Ekvall and Weidema (2004) suggest to consider only the additional impacts due to the additional supply of the required product for the CO_2-based process (black arrows in Figure 3.3). In this case, the environmental impacts of the co-product that is not considered in the LCA study are not affected from the additional supply of the other co-product (gray arrows in Figure 3.3). Considering only the consequences, the supply of co-products by combined production can be considered as a mono-functional processes. Thus, neither system expansion nor avoided burden is required. In this thesis, the following product from combined production is considered:

- CO_2 captured from flue gases (base case for CO_2 supply in this thesis).

If CO_2 supply from flue gases is considered as combined production, only the additional energy demand for the CO_2 capture process and the reduced CO_2 emissions must be considered. In von der Assen et al. (2016) and Reiter and Lindorfer (2015), the CO_2 supply from flue gases was considered as combined production. The consid-

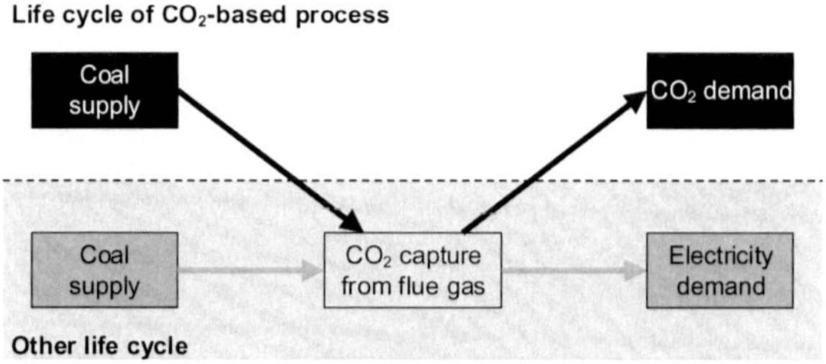

Figure 3.3: Consequences of multi-functional processes with combined production illustrated for CO_2 capture from flue gases. The black arrows indicate increased supply or demand. The gray arrows indicate constant supply or demand.

eration of CO_2 supply as combined production seems in particular appropriate if an existing process is retro-fitted with a CO_2 capture unit (von der Assen et al., 2016).

For CO_2 supply, a drawback of excluding the CO_2 source is that it leads to negative global warming impacts for CO_2 supply. This can lead to misinterpretation such as that CO_2 utilization is a net greenhouse gas sink. However, actually, the global warming impacts are only lower compared to the chosen benchmark process. Thus, in Chapter 5 the CO_2 source is included. Furthermore, for most CO_2 capture processes, data is only available that includes the production of the main product.

Joint production

For the joint production of co-products, the ratio between co-products is fixed. In other words, the increased supply of 1 co-product always also increases the supply of the other co-product. For the identification of affected processes in other life cycles, it is important to distinguish whether the utilization of the required co-product leads to an increased production of the multi-functional process or not (Ekvall and Weidema, 2004). Following Jung et al. (2013), co-products that increase production are defined in this thesis as main products and co-products that do not increase production as by-products.

Use of main product. In this case, the CO_2-based processes (or fossil-based processes) use a main product of a multi-functional process that increases its production. The supply of the by-product is also increased. If the increased supply of the by-product is not used for the CO_2-based process (or fossil-based process), other life cycles are affected (Figure 3.4). According to Ekvall and Weidema (2004), the utilization of the by-product in other life cycles should be included in the system boundaries. In this thesis, the following main product from joint production is considered:

- hydrogen from water electrolysis.

For hydrogen supply by water electrolysis, the utilization of the by-product oxygen is considered as a scenario in Chapter 5. Here, the substitution of conventional oxygen supply is considered (avoided burden). The utilization of by-products is not considered as base case in this thesis, because utilization cannot always be achieved.

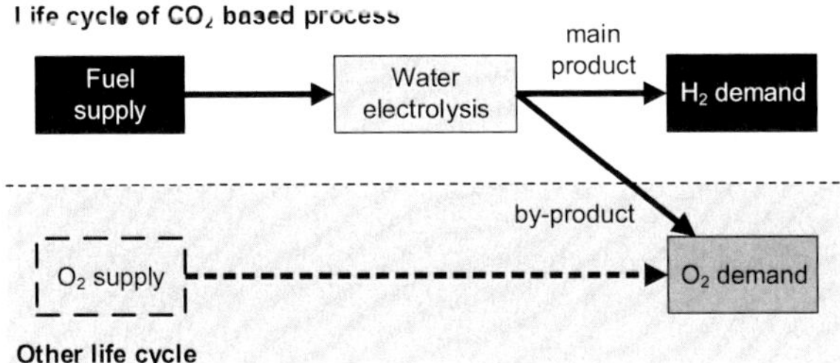

Figure 3.4: Consequences of multi-functional processes where main product is used in life cycle investigated illustrated for water electrolysis. The solid black arrows indicate increased supply or demand. The dashed black arrows indicate decreased supply or demand.

Use of by-product. Ekvall and Weidema (2004) assume that the additional demand for a by-product from joint production will not increase the supply of the by-product. Instead, the additional utilization of the by-product will lead to a reduced utilization of the by-product in other life cycles (Figure 3.5). In this thesis, the following by-product from joint production is considered:

- hydrogen from chemical processes such as chlorine electrolysis.

The supply of hydrogen from chlorine electrolysis is considered in Chapter 4. There, a case is analyzed where the by-product hydrogen from chlorine electrolysis is used as fuel for heating purposes (Jung et al., 2014). In this case, an alternative source for the heat (e.g., heat from natural gas) is required if hydrogen is used for CO_2-based processes.

Supply of by-product. If the CO_2-based processes (or fossil-based processes) produce co-products that are not used for CO_2-based processes (or fossil-based processes),

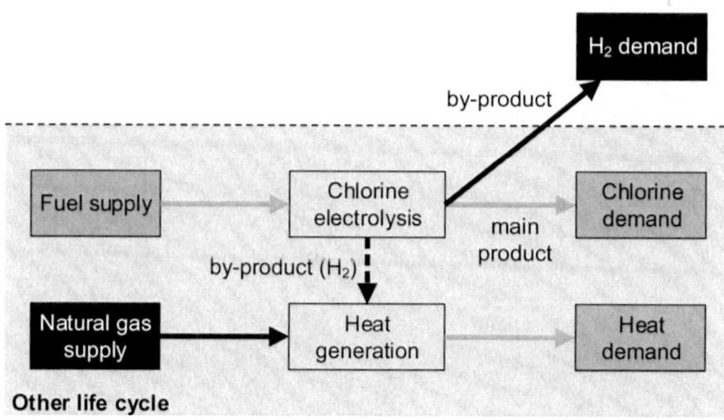

Figure 3.5. Consequences of multi-functional processes where by-product is used in life cycle investigated illustrated for chlorine electrolysis. The solid black arrows indicate increased supply or demand. The dashed black arrows indicate decreased supply or demand. The gray arrows indicate constant supply or demand.

other life cycles are affected (Figure 3.6). In this thesis, the following by-product is considered:

- heat from CO_2-based processes (carbon monoxide and methane production).

The utilization of the by-product heat is considered as a scenario in Chapter 5. Here, the substitution of natural-gas-based heat supply is considered (avoided burden).

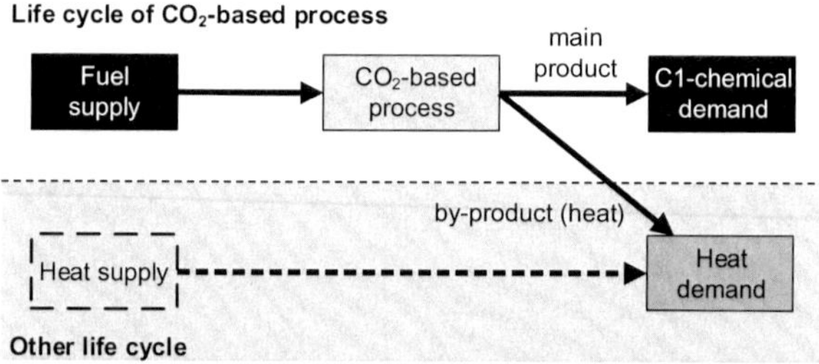

Figure 3.6: Consequences of by-product supply from life cycle investigated illustrated for CO_2-based processes. The solid black arrows indicate increased supply or demand. The dashed black arrows indicate decreased supply or demand.

Supply of main product. If the CO_2-based or fossil-based processes produce main products as co-products that are not used for CO_2-based or fossil-based processes, other life cycles are affected (Figure 3.7). In this thesis, the following main product is considered:

- hydrogen from fossil-based carbon monoxide and formic acid production.

In this case, system expansion is used. Thus, the corresponding CO_2-based process must also supply hydrogen.

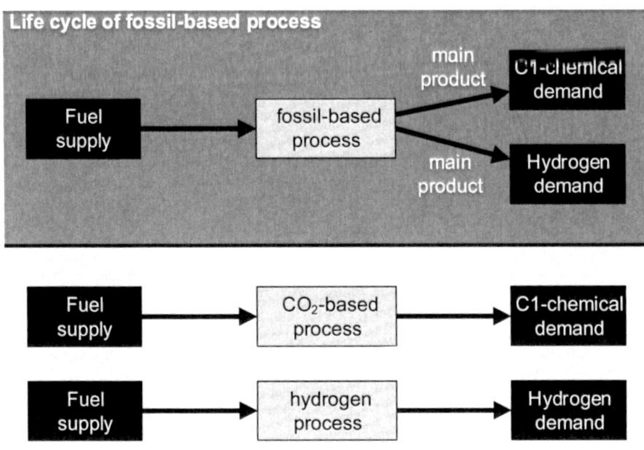

Figure 3.7: Consequences of main product supply from life cycle investigated illustrated for fossil-based processes. The black arrows indicate increased supply or demand.

Special case: CO_2 supply. According to the goal to provide a system-wide perspective, it is also taken into account that the supply of CO_2 affects the production of the co-produced product, e.g., electricity (Figure 3.8). The assumption that the CO_2 supply affects the production of the co-product seems for example appropriate if a new process with CO_2 capture is installed.

In the special case, the CO_2-based processes produce 2 main products (C1-chemical and for example electricity). To make the CO_2-based processes comparable to the fossil-based processes, system expansion is applied. This approach is investigated in Chapter 5. In Chapter 5, fossil-based power plants without CO_2-capture are used for system expansion. In Appendix D, several other power plants (e.g., wind power) are considered for system expansion. The consideration of an affected co-product (e.g., electricity) for CO_2 supply corresponds to the approach in van der Giesen et al. (2014) (see CO_2 supply in Section 2.3).

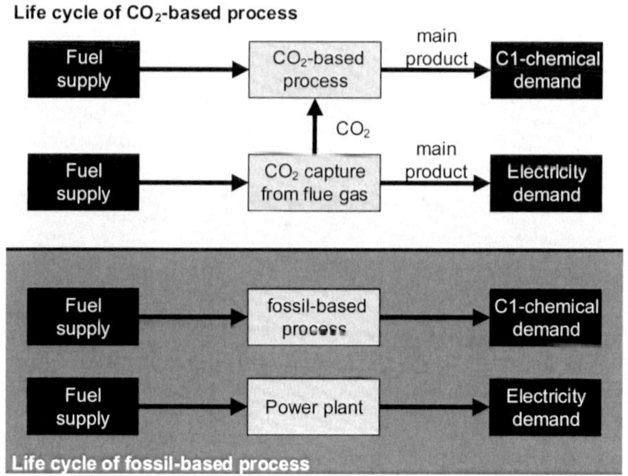

Figure 3.8: Consequences of CO_2 supply. The black arrows indicate increased supply or demand.

3.4 Use of renewable inputs

CO_2-based processes are often discussed as option to integrate renewable energy into the chemical sector (Perathoner and Centi, 2014). However, currently, it is not possible to satisfy the total energy demand with renewable energy. In 2013, renewable energies accounted for about 19 % of the global final energy demand (REN21, 2015). Thus, a close look at the consequences is required if renewable inputs are used for new processes. In this thesis, 3 potential consequences are distinguished (Figure 3.9):

- Case I: If the supply of renewable inputs is increased in sufficient amounts such that even the demand of the proposed technologies is satisfied, no consequence occur in other life cycles.
- Case II: If the supply of renewable inputs is increased but not large enough to satisfy the demand of all proposed technologies, consequences occur in the life cycle of other proposed technologies.

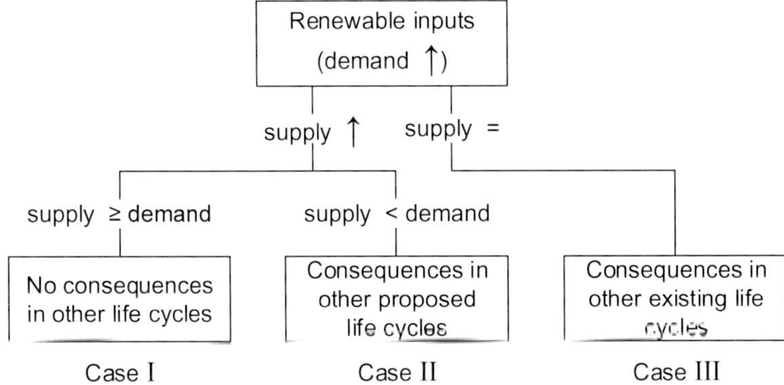

Figure 3.9: Potential consequences for the utilization of renewable inputs in other life cycles

- Case III: If the supply of the renewable inputs is constant, consequences occur in the life cycle of other existing technologies.

In this thesis, an approach is presented for the assessment of CO_2-based processes that takes into account all 3 potential consequences. For this purpose, the **utilization of the renewable inputs is considered in the functional unit**. In other words, the integration of renewable energy is considered as function of the CO_2-based processes. The advantages of considering the utilization of renewable inputs in the functional unit is presented in the following section.

In this thesis, the following renewable inputs are considered:

- electricity supply by renewable energy sources and
- biomass supply.

3.5 Maximum environmental impact reductions for CO_2-based processes

In this thesis, the consequences due to the introduction of CO_2-based processes are expressed as environmental impact reductions (Figure 3.10). Figure 3.10 shows how the interactions of other life cycles with the CO_2-based processes are taken into account in this thesis. In this section, the determination and application is presented first for the *environmental impact reductions* and then for the *maximum environmental impact reductions*.

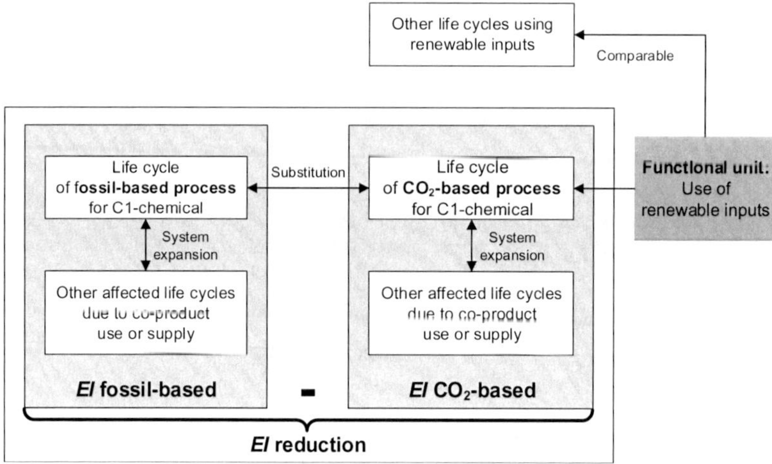

Figure 3.10: Interactions caused by CO_2-based processes considered in this thesis

3.5.1 Environmental impact reductions

The environmental impact reductions ($EI_{\text{reduction}}$) are the difference of the environmental impact of the fossil-based process ($EI_{\text{fossil-based}}$) and the CO_2-based process ($EI_{CO_2\text{-based}}$) (Equation 3.1). The environmental impact reductions are determined in this thesis for the following "function" of the CO_2-based processes: use of renewable inputs.

$$EI_{\text{reduction}} = EI_{\text{fossil-based}} - EI_{CO_2\text{-based}} \tag{3.1}$$

As shown in Figure 3.10, the interactions of CO_2-based processes with multi-functional supply processes and other competing processes for C1-chemicals are considered in the environmental impact reductions. The interactions with other processes using renewable inputs are not considered. Rather, the environmental impact reductions can be used directly to compare processes using renewable inputs.

Depending on the 3 potential consequences for the utilization of renewable inputs (cf. Figure 3.9), the CO_2-based processes are environmentally beneficial compared to fossil-based processes if the following conditions are satisfied:

Case I: The environmental impact of the CO_2-based process is lower than the environmental impact of the fossil-based process and thus the environmental impact reduction ($EI_{\text{reduction, CO}_2\text{-based}}$) according to Equation 3.1 is greater than 0.

$$EI_{\text{reduction, CO}_2\text{-based}} > 0 \tag{3.2}$$

Case II: In this case, an environmental impact reduction ($EI_{\text{reduction, CO}_2\text{-based}}$) greater than 0 is not sufficient for environmentally beneficial CO_2-based processes. Additionally, the analyzed CO_2-based process must achieve highest environmental impact reductions among other proposed processes that could use the renewable input.

$$EI_{\text{reduction, CO}_2\text{-based}} > 0 \text{ , and} \tag{3.2}$$

$$EI_{\text{reduction, CO}_2\text{-based}} > \max_i EI_{\text{reduction}, i} \tag{3.3}$$

with i – other proposed process that use renewable inputs

Case III: In this case, the utilization of renewable inputs for the CO_2-based process leads to a reduced availability of renewable inputs for other processes. The utilization of alternative inputs for the other processes changes the environmental impacts ($\Delta EI_{\text{other process}}$). Usually, the environmental impacts will increase ($\Delta EI_{\text{other process}} > 0$). Here, the utilization of renewable inputs for CO_2-based processes is environmentally beneficial from a system-wide perspective if the environmental impact reduction of the CO_2-based process ($EI_{\text{reduction, CO}_2\text{-based}}$) is greater than the change of environmental impacts in the other process.

$$EI_{\text{reduction, CO}_2\text{-based}} > \Delta EI_{\text{other process}} \tag{3.4}$$

3.5.2 Maximum environmental impact reductions

A drawback of the environmental impact reduction is that it depends on the environmental impact of the renewable input. Thus, if the utilization of several renewable inputs should be considered, the calculation of the environmental impact reductions must be repeated. To avoid this dependence, the environmental impact of renewable inputs is not considered for CO_2-based processes ($EI_{\text{CO}_2\text{-based, without renewable input}}$) in this thesis for the determination of environmental impact reductions. The resulting environmental impact reductions are denoted maximum environmental impact reductions ($EI^{\text{max}}_{\text{reduction}}$):

$$EI^{\text{max}}_{\text{reduction}} = EI_{\text{fossil-based}} - EI_{\text{CO}_2\text{-based, without renewable input}}. \tag{3.5}$$

The maximum environmental impact reduction ($EI^{max}_{reduction}$) is the sum of the environmental impact reduction ($EI_{reduction}$) and the environmental impact of renewable input supply ($EI_{renewable\ input\ supply}$):

$$EI^{max}_{reduction} = EI_{reduction} + EI_{renewable\ input\ supply}. \quad (3.6)$$

If the maximum environmental impact reduction is used for the assessment of CO_2-based processes, the necessary condition for environmentally beneficial CO_2-based processes is that the maximum environmental impact reduction is greater than the environmental impact of renewable input supply. In other words, the condition in Equation 3.2 for environmentally beneficial CO_2- based processes changes to:

$$EI^{max}_{reduction} > EI_{renewable\ input\ supply}. \quad (3.7)$$

The other conditions (Equations 3.3 and 3.4) are also valid for maximum environmental impact reductions.

In this thesis, the maximum environmental impact reduction is used for the environmental analysis of CO_2-based C1-chemicals using renewable inputs. The maximum environmental impact reduction allows to assess all previously presented cases for the supply of renewable inputs (Figure 3.9).

In Chapter 4 and 5, the maximum environmental impact reduction is predominantly used to analyze the case where the use of renewable inputs causes no consequences in other life cycles (Case I). In this case, the maximum environmental impact reduction is used as threshold value for the supply of renewable inputs (Equation 3.6). In other words, the maximum environmental impact reduction is used to identify renewable input supply processes that lead to environmentally beneficial CO_2-based processes. In Chapter 4, the supply of hydrogen is analyzed and in Chapter 5 the supply of electricity that is used in an electrolysis to produce hydrogen. In Chapter 4 and 5, the maximum environmental impact reduction is also used to identify non-renewable inputs that lead to environmentally beneficial CO_2-based processes.

In Chapter 6 and 7, the maximum environmental impact reduction is used to compare the environmental impact reduction of CO_2-based processes with environmental impact reductions of other proposed processes that can use renewable inputs (Case II for renewable inputs).

The comparison of CO_2-based processes with existing life cycles that use renewable inputs (Case III for renewable inputs) is not within the scope of this thesis, because the share of renewable inputs is expected to increase (REN21, 2015).

Chapter 4

CO_2-to-Chemical – Deriving the environmental impact reductions

In this chapter, the CO_2-based and fossil-based processes considered for formic acid, carbon monoxide, methanol and methane are presented (Section 4.2). In contrast to the CO_2 conversion reactions already presented in Section 2.1.2, the conversion processes include the purification of product mixes to the pure C1-chemicals. In Section 4.3, the maximum environmental impact reduction is introduced for CO_2-based processes. The reduction of environmental impacts result from the substitution of fossil-based processes. In Section 4.4, the maximum environmental impact reductions are compared for all considered CO_2-based process. Finally, conclusions are drawn in Section 4.5.

Major parts of this chapter are reproduced with permission from The Royal Society of Chemistry from:

> Sternberg, A., Jens, M.C. and Bardow, A. (2017). Life Cycle Assessment of CO_2-based C1-chemicals, *Green Chemistry*, 19(9):2244-2259.

4.1 Goal and scope definition

Goal. In this chapter, a comparative life cycle assessment is presented for CO_2-based production of formic acid, carbon monoxide, methanol and methane. CO_2 is converted using hydrogenation. Thus, the availability of hydrogen and the corresponding environmental impacts are central to the CO_2-based processes. Current LCA studies show that CO_2-based processes only reduce environmental impacts compared to fossil-based processes if hydrogen is supplied by renewable energies. If hydrogen is supplied by conventional steam-methane-reforming, the CO_2-based processes increase environmental impacts compared to fossil-based processes (Aresta et al., 2002; von der Assen et al., 2013; Reiter and Lindorfer, 2015). The main goals of this chapter are therefore:

I *Comparison of the maximum environmental impact reduction for all CO_2-based processes.* The maximum environmental impact reduction allows to compare all CO_2-based processes assuming limited hydrogen.

II *Determination of environmental threshold values for hydrogen supply processes.* The environmental threshold values are the allowable impact of hydrogen supply processes to still lead to environmentally beneficial CO_2-based processes compared to their corresponding fossil-based processes.

Functional unit. In this chapter, the following functional unit is chosen:

- Utilization of 1 kg hydrogen for CO_2-based processes.

This functional unit allows to analyze all hydrogen supply processes and not only hydrogen supplied by an electrolysis unit. Nonetheless, the order for the maximum environmental impact reductions also applies for renewable electricity because all processes use the same amount of hydrogen. Consequently, all processes require the same amount of renewable electricity if the efficiency of the electrolysis is equal.

System boundaries. The maximum environmental impact reductions are derived by comparing CO_2-based processes for C1-chemicals to their corresponding fossil-based processes. For the comparison, the fossil-based processes must be adjusted to produce the same amount of C1-chemicals as the CO_2-based processes by using 1 kg hydrogen. For methane, the fossil-based process is adjusted to produce the same amount of energy based on the lower heating value (LHV) as the CO_2-based process. Methane is compared on an energy basis rather than mass, because CO_2-based methane is compared to natural gas with different compositions and thus different lower heating values. For each C1-chemicals, the amount of product yielded from

1 kg hydrogen depends on the process concept of the CO_2-based process. In the Appendix, the amount of C1-chemicals yielded from 1 kg hydrogen is presented in Table C.1.

In the comparison of CO_2-based to fossil-based C1-chemicals, the use phase and end-of-life phase are identical and are therefore not considered. In this chapter, the cradle-to-gate approach is used considering only the upstream processes (except hydrogen supply) for both CO_2-based and fossil-bases processes for C1-chemicals (Figure 4.1).

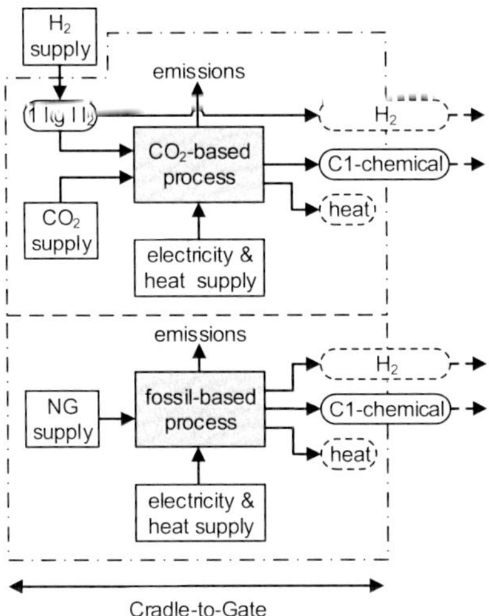

Figure 4.1: System boundaries for CO_2-based processes (top) and fossil-based processes (bottom). Note that the necessary supply processes vary with each C1-chemical and not all C1-chemicals produce hydrogen as co-product. NG – natural gas

The cradle-to-gate approach requires that each CO_2-based and the corresponding fossil-based process supply the same product (C1-chemical). However, besides the C1-chemical, the CO_2-based and corresponding fossil-based process can also produce non-common products. E.g., hydrogen is co-produced by the fossil-based carbon monoxide production. In this case, the CO_2-based processes must be adjusted to produce the same amount of hydrogen per carbon monoxide as the fossil-based process (System expansion, cf. Section 3.3). Since carbon monoxide is a precursor of the conventional formic acid process, system expansion is used for production of both carbon monoxide (products: carbon monoxide and hydrogen) and formic acid (products: formic acid and hydrogen). Consequently, for CO_2-based production of carbon monoxide and formic acid, not all of the total amount of 1 kg hydrogen is used in the CO_2-based process to produce the corresponding C1-chemical. Rather, a share of the hydrogen is used to substitute hydrogen that is co-produced by the fossil-based processes (see also Figure 2.1 and Table C.1 in the Appendix).

Environmental indicators. In this chapter, impact reductions are derived for global warming (GW) and fossil depletion (FD). These impact categories are chosen, because they reflect the major motivation for installing renewable energies. The environmental impacts are determined according to ReCiPe 1.08 Midpoint (Hierarchist) (Goedkoop et al., 2009).

4.2 CO_2-based and fossil-based processes for C1-chemicals

In this section, processes for C1-chemicals are briefly presented. In contrast to the CO_2 conversion reactions presented in Section 2.1.2, the process routes also consider process engineering aspects such as separation of products. In the following, CO_2-based and conventional process routes are presented for formic acid, carbon monoxide, methanol and methane. The inputs and outputs of the considered process routes are basis for the life cycle inventory in this thesis.

4.2.1 CO_2-based processes

This section gives an overview of CO_2-based process concepts for C1-chemicals. The processes are presented in the order of increasing hydrogen demand per mole of CO_2 used. For each C1-chemical, the key process steps are mentioned. Flowsheets and process data from the process simulations considered for the life cycle assessment are presented in the Appendix (Figures A.1–A.9). For the environmental analysis in the following sections, all results are based on the supply of hydrogen and CO_2 at 1 bar to the CO_2-based processes.

Formic acid. In this thesis, 2 process simulations are considered from Pérez-Fortes et al. (2016) and Jens et al. (2017) for the patented BASF process concept using the amine trihexylamine and the solvents methanol/water (Schaub et al., 2012). Pérez-Fortes et al. (2016) analyze the reaction of CO_2 and hydrogen to formic acid at 93 °C and 105 bar. After the reaction, stripping is performed at 3 bar before formic acid (purity: 85 wt%) is recovered at the top of the reactive distillation column. In the process simulation of Jens et al. (2017) the CO_2 and hydrogen react at 50 °C and 94 bar. Subsequently, methanol, water and residual CO are removed by distillation. Then, the amine-formic-acid adduct is decomposed in a reactive distillation at 200 mbar and 180 °C to liberate the pure formic acid, which is recovered at the top of the reactive distillation column.

Carbon monoxide via reverse water-gas-shift (rWGS). The rWGS process was analyzed in the project CO_2RRECT (2014). Hydrogen, CO_2 and a hydrogen-rich recycle stream are fed into the reactor. At ambient pressure and over 900 °C, the equilibrium conversion of CO_2 to CO reaches about 90 % (Kaiser et al., 2013). After the reaction, the raw gas contains CO, CO_2, water, hydrogen and traces of methane. To obtain pure CO, syngas treatment is required. First, water is separated by cooling down the hot raw gas. In the next step, CO_2 is removed by amine scrubbing and recycled to the reactor. Then, hydrogen and methane are separated from the raw gas by liquid methane scrubbing in the so-called cold-box. The outputs of the liquid methane scrubbing are pure CO, a hydrogen-rich stream and a purge gas stream containing methane and CO. The hydrogen-rich stream is recycled to the reactor (CO_2RRECT, 2014). In CO_2RRECT (2014), an electrical heating concept is considered for the reactor. In this thesis, additionally heating by steam is considered. In this chapter, the purge gas stream is not valorized. A utilization of the purge gas stream for heat supply is considered in Chapter 5.

Carbon monoxide via dry reforming of methane (DRM). A process concept for the production of CO via DRM was also investigated in the project CO_2RRECT (2014). CH_4 and CO_2 are fed into the electric heated reactor. To avoid the production of coke, temperatures over 1000 °C are required. At ambient pressure and 1000 °C, the equilibrium conversion of CO_2 is about 95 % (Jafarbegloo et al., 2015). After the reaction, the raw gas contains CO, CO_2, water, hydrogen and small amounts of methane. Both products hydrogen and CO can be separated by a similar raw gas treatment as for the rWGS reaction. However, for DRM, in the last step, an additional pressure swing adsorption (PSA) is required to further purify the hydrogen-rich stream leaving the liquid methane scrubbing. In the DRM process, only CO_2 is recycled to the reactor. As for the rWGS process, data is used from the project CO_2RRECT (2014), where the reactor is heated electrically. For the environmental assessment in this thesis, 2 concepts to heat the reactor are considered: electrical heating and by steam.

Methanol. Rihko-Struckmann et al. (2010) presented a process simulation for the CO_2-based methanol production. Hydrogen and CO_2 are fed into a reactor cascade consisting of 4 adiabatic reactors at 220 °C and 50 bar. Between the reactors, the gas is cooled. After the reactor, the methanol is separated from unreacted components such as CO, CO_2 and hydrogen by a flash unit. The gaseous components are then recycled to the reactor. The methanol-rich stream is further purified in a methanol distillation. A similar process simulation was presented by Van-Dal and Bouallou (2013) with different reaction conditions (210 °C and 78 bar). The process simulation presented by Kiss et al. (2016) considers an additional stripping unit to increase the CO_2 conversion. Here, the hydrogen feed (saturated with water) is contacted with the methanol-water mixture to recover unconverted CO/CO_2. The reaction is carried out at 250 °C and 50 bar. The process simulations presented by Van-Dal and Bouallou (2013) and Rihko-Struckmann et al. (2010) supply excess heat. However, in this thesis the utilization of excess heat is not considered. For the process simulation presented by Rihko-Struckmann et al. (2010) the excess heat is less than 5 % compared to the LHV of the methanol output. Van-Dal and Bouallou (2013) did not explicitly state the amount of excess heat. The process simulation presented by Kiss et al. (2016) requires additional heat supply due to the utilization of wet hydrogen.

Methane. In this thesis, 2 process simulations are considered from Müller et al. (2011) and Jean et al. (2014). The reaction conditions in the proposed process are in the range of 280–300 °C and 5–8 bar. Müller et al. (2011) present a process concept where only water removal is required before the so-called substitute natural gas (SNG) can be fed into the natural gas grid. The SNG contains 88 vol% methane, 6 vol% hydrogen and 6 vol% CO_2. In the process of Jean et al. (2014), the additional removal of CO_2 and hydrogen is considered before the water is removed and the SNG is fed into the natural gas grid. The SNG contains 96.8 vol% methane, 1.6 vol% hydrogen and 1.6 vol% CO_2. In this chapter, the utilization of the excess heat from the reaction is not considered. A utilization of the excess heat is considered in Chapter 5.

4.2.2 Fossil-based processes

Natural gas. The fossil counterpart of CO_2-based methane is natural gas. Raw natural gas is a naturally occurring gas mixture that is recovered from underground deposits. Raw natural gas mainly contains methane (75–99 vol%) (Hammer et al., 2000). Before natural gas is supplied to consumers, water and impurities such as sulfur and mercury are removed. The natural gas is compared to CO_2-based methane on a lower heating value (LHV) basis. For the life cycle assessment, the natural gas supply with highest and lowest environmental impacts in the EU-27 are considered.

Carbon monoxide. Pure carbon monoxide is typically produced by separating it from syngas. Thus, carbon monoxide is co-produced with hydrogen and/or syngas (Ledon, 1986). The industrial standard process for syngas production is steam-methane-reforming (SMR) (Chen et al., 2005):

$$CH_4 + H_2O \rightleftharpoons CO + 3H_2. \tag{4.1}$$

For carbon monoxide production, the SMR reaction is typically carried out at 850–900 °C and 10–30 bar (Ledon, 1986). After the reaction, the raw gas contains carbon monoxide, hydrogen, CO_2, water and methane. The molar H_2/CO ratio in the raw gas typically ranges from 4 to 5 if only methane and water are used as feed to the reactor. The molar H_2/CO ratio can be decreased to about 3 if CO_2 from the raw gas is recycled to the reactor. In this chapter, 2 fossil-based processes are considered for carbon monoxide production with the following molar H_2/CO ratios: 3 (CO_2RRECT, 2014) and 4.4 (Baltrusaitis and Luyben, 2015).

Methanol. Conventional methanol production consists of the following 2 reaction steps:

i) Production of syngas. The syngas for methanol production mainly contains hydrogen, carbon monoxide and CO_2. The syngas composition strongly affects the methanol yield. The composition of the syngas is described by the so-called stoichiometric number (SN):

$$SN = \frac{y_{H_2} - y_{CO_2}}{y_{CO} + y_{CO_2}}, \tag{4.2}$$

where y is the concentration in volume percent. A stoichiometric number of $SN = 2$ corresponds to the stoichiometric demand of reactants for methanol synthesis. For methanol production, the majority of syngas is produced by:

- steam-methane-reforming: The stoichiometric number (SN) of the syngas from steam-methane-reforming is typically between 2.6 and 2.9 (Fiedler et al., 2000), i.e., the syngas contains too much hydrogen for methanol synthesis. After the methanol synthesis, a hydrogen-rich purge gas is separated from the methanol stream. The hydrogen-rich purge stream is usually used for heating purposes.
- combined reforming: In the combined reforming process, 40–60 % of natural gas feed is reformed by steam-methane-reforming (SMR). The remaining 60–40 % of natural gas feed is reformed with oxygen and the syngas of the SMR in an autothermal reformer. The resulting syngas achieves a stoichiometric number of $SN \approx 2$ (Fiedler et al., 2000).

ii) Methanol synthesis. For the methanol synthesis, the syngas is converted to methanol according to the following reactions:

$$CO + 2H_2 \rightleftharpoons CH_3OH, \tag{4.3}$$

$$CO_2 + 3H_2 \rightleftharpoons CH_3OH + H_2O. \tag{4.4}$$

The syngas is converted to methanol at about 200–300 °C and 50–100 bar (Fiedler et al., 2000). In the reactor, about 50 % of syngas are converted to methanol. The unreacted syngas is recycled to the reactor after methanol and water are condensed out. The pure methanol is separated from water in a distillation column.

For fossil-based methanol production, both a high-efficiency and a low-efficiency process are considered according to ecoinvent (2007a). The high-efficient process corresponds to syngas production by combined reforming and the low-efficient process corresponds to syngas production by steam-methane-reforming.

Formic acid. Conventionally, formic acid is primarily produced by the hydrolysis of methyl formate (Hietala et al., 2016). The methyl formate hydrolysis process consists of the following reaction steps:

i) Production of carbon monoxide. For the carbon monoxide production, the above described fossil-based processes are considered with both H_2/CO ratios (3 and 4.4).

ii) Carbonylation of methanol. Methanol is carbonylated in the liquid phase with carbon monoxide at about 45 bar and 80 °C (Hietala et al., 2016):

$$CH_3OH + CO \rightleftharpoons HCOOCH_3 \tag{4.5}$$

iii) Hydrolysis of methyl formate. In the equimolar conversion of methyl formate and water, only about 30 % of methyl formate is converted. The hydroloysis is equilibrium limited, thus the conversion can be increased by a large excess of either methyl formate or water.

$$HCOOCH_3 + H_2O \rightleftharpoons HCOOH + CH_3OH \tag{4.6}$$

For the purification of the resulting product mixture mainly 2 concepts exist (ecoinvent, 2007c):

a) Utilization of 2 distillation columns and

b) Extraction using a secondary amide to extract formic acid from the aqueous solution, before final purification with distillation.

For carbonylation of methanol (step ii) and hydrolysis of methyl formate (step iii) of the fossil-based formic acid production, only 1 process concept is considered due to missing alternative data. However, for the supply of carbon monoxide (step i), both fossil-based processes (H_2/CO ratio = 3 and 4.4) are considered.

4.2.3 Feedstock and utility supply processes

The process demands for feedstock and utilities presented in the previous subsections are used to determine the environmental impacts of CO_2-based and fossil-based processes for C1-chemicals. In this section, the selected supply processes are presented for feedstock and utilities (heat and electricity). The environmental impacts of the supply processes are taken from the LCA-database GaBi ts (2016).

Natural gas supply. The EU-27 natural gas mix is applied.

Heat supply. Heat is supplied by steam from natural gas with an efficiency of 90 % (see Table B.1).

Electricity supply. A forecasted electricity mix of the EU-27 for 2020 is considered (see Table B.1).

CO_2 supply. A range of supply processes is considered, because currently no standard CO_2 supply process is established for large-scale CO_2 utilization in the chemical industry. The CO_2 supply processes are categorized into the following 2 groups:

CO_2 capture from flue gas. CO_2 is part of the flue gas of all fossil-fired power plants and many industrial processes, such as chemicals and cement production. For the supply of captured CO_2, von der Assen et al. (2016) recently determined the impacts for global warming (−0.99 to −0.42 kg CO_2-eq/ kg CO_2) and fossil depletion (0.00–0.19 kg oil-eq/kg CO_2). The environmental impact of each CO_2 capture process is based on a comparison to a corresponding process without CO_2 capture. For example, a coal-fired power plant with CO_2 capture is compared to a coal-fired power plant without CO_2 capture. The global warming impact is negative, because the process with CO_2 capture has a lower global warming impact than the process without CO_2 capture (see also Table A.1).

Air capture. The air capture process removes CO_2 from the air and supplies captured CO_2. The CO_2 concentration in air is about 400 ppm, which is significantly lower than in flue gases from industrial processes. Thus, the energy demand for CO_2 capture from air is higher than from flue gas. The higher energy demand results in higher global warming and fossil depletion impacts than CO_2 capture from flue gases if fossil fuels are used for energy supply. In this case, the global warming impacts for air capture range from −0.58 to −0.42 kg CO_2-eq/ kg CO_2. The fossil depletion impacts range from 0.16 to 0.24 kg oil-eq/ kg CO_2 (von der Assen et al., 2016). If renewable energy is used for energy supply, a global warming and fossil depletion impact can be achieved close to −1 kg CO_2-eq/ kg CO_2 and 0 kg oil-eq/ kg CO_2, respectively.

Thus, to include all CO_2 sources analyzed by von der Assen et al. (2016), the following ranges for CO_2 supply are applied in this chapter:

- Global warming impact: -0.42 to -0.99 kg CO_2-eq/ kg CO_2,
- Fossil depletion impact: 0 to 0.24 kg oil-eq/ kg CO_2.

Note that the CO_2 capture processes presented by von der Assen et al. (2016) consider the compression of CO_2 to 100 bar, which is not required for the CO_2-based processes. However, since the best-case values for global warming impact (-1 kg CO_2-eq/ kg CO_2) and fossil depletion impact (0 kg oil-eq/ kg CO_2) are almost achieved, the ranges from von der Assen et al. (2016) are used.

In the Appendix (Figure C.6), the full range of global warming impacts for CO_2 supply are considered (GW = 0 to -1 kg CO_2-eq/ kg CO_2). A global warming impact of zero for CO_2 supply corresponds to the utilization of CO_2 that otherwise would be store in the underground (CCS).

Hydrogen supply. In the following, 3 hydrogen supply processes are presented to relate the maximum environmental impact reductions to existing and proposed hydrogen supply processes. However, note that the hydrogen supply processes are not required to calculate the the maximum environmental impact reductions.

Steam-methane-reforming (SMR). The process currently most used for hydrogen production is steam-methane-reforming with a subsequent water-gas-shift reaction (Baufumé et al., 2013). For Germany, the global warming and fossil depletion impact of hydrogen supply by SMR is 10.6 kg CO_2-eq and 4.5 kg oil-eq per kg hydrogen (GaBi ts (2016), see also Table B.4). This process requires about 187 MJ of natural gas (LHV) per kg hydrogen (120 MJ based on LHV). The corresponding net energy ratio is 0.64. The required amount of natural gas already includes a credit for natural gas because steam-methane-reforming co-produces electricity from excess heat. The credit for natural gas corresponds to the amount of natural gas that would be required to produce the amount of electricity.

Hydrogen from thermal utilization. Hydrogen is a by-product of many chemical processes. In large interconnected chemical plants, the by-product hydrogen is used as feedstock for other chemical processes. However, some processes, e.g., chlor-alkali electrolysis produce the by-product hydrogen in excess. Then, hydrogen is usually used as fuel for heating purposes (Jung et al., 2014). This hydrogen could be used as feedstock for CO_2 hydrogenation. However, using by-product hydrogen as feedstock requires an alternative source for the heat supplied before by hydrogen, e.g., heat from natural gas. In this case, the environmental impact of supplying 1 kg hydrogen equals the environmental impact of supplying 33.3 kWh heat (lower heating value of hydrogen) from natural gas. In this case, the global warming and fossil depletion impact of hydrogen supply is 7.9 kg CO_2-eq and 3.1 kg oil-eq per kg hydrogen.

Water electrolysis. The environmental impact of hydrogen from water electrolysis predominantly depends on the efficiency of the electrolysis and the environmental impact of electricity employed. Considering a state-of-the art efficiency of 50 kWh per kg H_2 for a polymer electrolyte membrane (PEM) electrolysis with a pressure of operation of about 30 bar (Bertuccioli et al., 2014), the global warming impact for the supply of 1 kg hydrogen ranges from 0.4 kg CO_2-eq (wind electricity) to 18.5 kg CO_2-eq (grid mix EU-27 in 2020). The corresponding fossil depletion impact ranges from 0.12 to 4.95 kg oil-eq per kg H_2.

4.3 Environmental impacts of CO_2-based and fossil-based processes

In this section, the environmental impacts of CO_2-based processes are analyzed in detail and compared to fossil-based process. Figure 4.2 shows global warming impacts of CO_2-based and fossil-based processes for formic acid (4.2a), carbon monoxide (4.2b & 4.2c), methanol (4.2d) and methane (4.2e). For the CO_2-based processes, a breakdown of the global warming impacts is shown. In Figure 4.2, a global warming impact for CO_2 supply of -0.7 kg CO_2-eq/kg CO_2 is considered (mean of considered range for CO_2 supply). For the breakdown analysis, hydrogen supply by steam-methane-reforming is considered. The global warming impact of hydrogen supply by steam-methane-reforming is 10.6 kg CO_2-eq/kg H_2. For all C1-chemicals, at least 2 CO_2-based process concepts are considered. Consequently, a range for global warming impacts is shown in Figure 4.2 (error bars). The global warming impact due to hydrogen supply is equal for all considered CO_2-based processes, because all processes use exactly 1 kg hydrogen (functional unit). For the CO_2-based production of formic acid, methanol and methane, the yield from 1 kg hydrogen varies due to the different process concepts considered. Thus, a range is presented for the global warming impact of the fossil-based processes, which reflects the varying amount of C1-chemicals. In this section, the CO_2-based processes are benchmarked to the fossil-based processes with the highest global warming impacts, i.e., the worst case for fossil-based processes. For the maximum environmental impact reduction presented in the next sections, a worst case and best case is considered for fossil-based processes. The fossil depletion impacts of CO_2-based and fossil-based processes are show in the Appendix (Figure C.1).

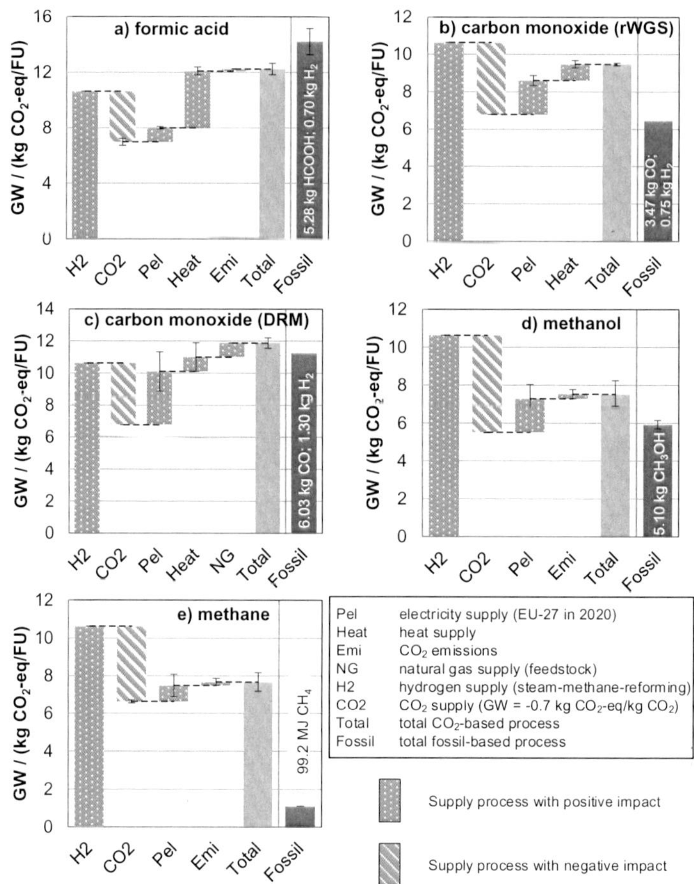

Figure 4.2: Breakdown of global warming (GW) impacts per functional unit (FU = use of 1 kg H_2) for CO_2-based processes. The error bars of the CO_2-based processes represent the range of considered process concepts. For fossil-based processes an upper bound of global warming impacts (worst case) is shown. The range of the fossil-based processes is due to different yields of C1-chemicals from the considered CO_2-based processes. The average yield is shown in the right bar.

Formic acid. The global warming impact of the CO_2-based process is about 12.2 kg CO_2-eq/FU. The majority of the global warming impact is due to hydrogen supply (10.6 kg CO_2-eq/FU). The global warming impact of CO_2 supply is about −3.6 kg CO_2-eq/FU. Heat and electricity supply account for about 5.1 kg CO_2-eq/FU. Compared to other CO_2-based processes, the contribution of heat supply is high, because of the complex product purification. While the environmental impacts of the 2 considered CO_2-based processes are almost equal, the amount of produced products differs by about 14 %. This is due to different overall hydrogen conversion rates for the CO_2-based formic acid process.

Due to the different amount of products for the CO_2-based processes, the global warming impact per FU of fossil-based formic acid (H_2/CO ratio = 3) ranges from 13.3 to 15.2 kg CO_2-eq/FU. The global warming impact per FU of the fossil-based process is highest among all considered fossil-based processes for C1-chemicals and higher than the global warming impact of the CO_2-based formic acid production. For fossil-based formic acid, process energy demand and conversion losses have a high share in total global warming impact (93 %). The global warming impact of the natural gas supply that is required as feedstock (86–99 MJ per FU) only accounts for about 7 % of the total global warming impacts.

Carbon monoxide (rWGS). The total global warming impact (9.5 kg CO_2-eq/FU) of the rWGS process is dominated by the global warming impact of the hydrogen supply (10.6 kg CO_2-eq/FU) and CO_2 supply (−3.8 kg CO_2-eq/FU). The impact due to heat and electricity supply is about 2.7 kg CO_2-eq/FU. The difference between the 2 considered rWGS process concepts is the heat supply for the reactor (electricity vs natural gas). Both processes yield the same amount of product per FU.

The global warming impact of the fossil-based process (H_2/CO ratio = 3) is about 6.5 kg CO_2-eq/FU. For fossil-based carbon monoxide, the global warming impacts due to heat and electricity supply account for about 83 %. The supply of natural gas used as feedstock (100 MJ per FU) accounts for 17 % of total global warming impacts.

Carbon monoxide (DRM). The global warming impact of the DRM process is 11.9 kg CO_2-eq/FU. Note that this value cannot be compared to the rWGS process because the DRM process produces almost the double amount of product (see Figure 4.2). The majority of the global warming impacts is due to the hydrogen supply (10.6 kg CO_2-eq/FU). Additionally, natural gas (0.9 kg CO_2-eq/FU) and CO_2 (−3.8 kg CO_2-eq/FU) are required as feedstocks. As for the rWGS process, 2 concepts for reactor heating (electricity and natural gas) are considered. Since the DRM reaction is more endothermic than the rWGS reaction, the range of the total global warming impact is larger for the DRM process than for the rWGS process.

The global warming impact of the fossil-based process (H_2/CO ratio = 3) is about 11.2 kg CO_2-eq/FU. In this case, the global warming impact of DRM is only 0.7 kg CO_2-eq/FU higher than the global warming impact of the fossil-based process.

Methanol. For CO_2-based methanol production, the total global warming impacts range from 6.9–8.2 kg CO_2-eq/FU. As for other CO_2-based processes, the majority of the global warming impact is due to hydrogen supply (10.6 kg CO_2-eq/FU). For methanol production, the contribution of CO_2 supply (−5.1 kg CO_2-eq/FU) to the total global warming impact is highest among all CO_2-based processes. The range of the total global warming impact is mainly due to different electricity demands in the proposed process concepts. The main difference in the process concepts is the pressure drop after the flash and the resulting electricity demand for the compressor of the recycling stream. Furthermore, the amount of produced products differs by up to 8 % for the considered CO_2-based processes.

Accordingly, the total global warming impact of the fossil-based methanol production ranges from 5.7–6.2 kg CO_2-eq/FU. The fossil-based processes require about 123–133 MJ natural gas per FU as feedstock. This value is higher than for fossil-based formic acid and carbon monoxide (rWGS) because the considered fossil-based processes for methanol do not produce hydrogen as co-product. Thus, the total amount of hydrogen for CO_2-based processes can be used for methanol production. Consequently, more natural gas is required for the corresponding amount of fossil-based methanol per FU. The supply of natural gas used as feedstock accounts for about 25 % of total global warming impacts.

Methane. The global warming impact of the CO_2-based process varies from 7.2–8.2 kg CO_2-eq/FU. The total global warming impact is mainly dominated by hydrogen supply (10.6 kg CO_2-eq/FU) and CO_2 supply (−4.0 kg CO_2-eq/FU). Due to different product purification concepts, the electricity demand of the 2 considered CO_2-based processes for methane ranges from 0.24 to 1.44 kg CO_2-eq/FU. For both process concepts, the amount of produced methane is almost equal (about 100 MJ per FU).

The global warming impact for the natural gas supply mix in the EU-27 is 1.12 kg CO_2-eq/FU. For fossil-based methane, the global warming impact originates to 100 % from the supply of natural gas.

In summary, the global warming impacts for all CO_2-based C1-chemicals range from 6.9–12.2 kg CO_2-eq per FU. The CO_2-based production of formic acid has the highest impacts and the CO_2-based production of methanol has the lowest impacts.

For the fossil-based C1-chemicals, the global warming impacts range from 1.1–15.2 kg CO_2-eq per FU. Here, the fossil-based production of formic acid has the highest impacts. The fossil-based production of methane has the lowest impacts.

The wide range of the global warming impacts of the fossil-based processes is primarily due to different process energy demands. For example, the fossil-based processes for formic acid, carbon monoxide (rWGS), and methane require almost the same amount of natural gas as feedstock per FU (86–100 MJ). This results in global warming impacts of about 0.95–1.12 kg CO_2-eq per FU. The fossil-based methane production does not have any further global warming impacts. The fossil-based CO production causes additional 5.4 kg CO_2-eq per FU through heat and electricity demand. The fossil-based formic acid production even causes additional 12–14 kg CO_2-eq per FU, because the fossil-based production of formic acid requires 3 reaction steps (when starting with CH_4) and a complex product purification (see also Figure C.2).

4.4 Maximum environmental impact reductions for CO_2-based processes

In Section 4.4.1, the determination of the maximum environmental impact reduction is presented. In Section 4.4.2, the interpretation of the maximum global warming impact reduction is explained in detail for CO_2-based formic acid production. In Section 4.4.3, the maximum environmental impact reductions are compared for all considered C1-chemicals.

4.4.1 Determination of maximum environmental impact reductions

If hydrogen is supplied by steam-methane-reforming as presented in Section 4.3, only CO_2-based formic acid production achieves lower global warming impacts than the fossil-based process in 2020. To determine suitable hydrogen supply processes that lead to environmentally beneficial CO_2-based processes for all C1-chemicals, maximum environmental impact reductions are determined for each C1-chemical.

The maximum environmental impact reductions can be determined by the following 3 steps:

1. The CO_2-based processes for C1-chemicals are modeled to determine input and output flows. The input and output flows are then used to determine the global warming impact ($GW_{CO_2\text{-based}}$) and fossil depletion impact ($FD_{CO_2\text{-based}}$) of the

CO_2-based processes for C1-chemicals according to

$$GW_{CO_2\text{-based}} = m_{CO_2} \cdot GW_{CO_2\text{ supply}} + Q \cdot GW_{\text{heat supply}} + P_{\text{el}} \cdot GW_{\text{electricity supply}} + m_{CO_2\text{ emissions}}, \tag{4.7}$$

and

$$FD_{CO_2\text{-based}} = m_{CO_2} \cdot FD_{CO_2\text{ supply}} + Q \cdot FD_{\text{heat supply}} + P_{\text{el}} \cdot FD_{\text{electricity supply}}. \tag{4.8}$$

Here, the input flows are heat (Q), grid power (P_{el}) and CO_2 (m_{CO_2}). The environmental impacts (GW_{supply} and FD_{supply}) for the supply process are explained in Section 4.2.3. For the determination of the maximum environmental impact reduction, the environmental impact of hydrogen supply is not considered.

2. The environmental impacts of fossil-based processes are determined in the same way as the CO_2-based processes:

$$GW_{\text{fossil-based}} = m_{\text{NG}} \cdot GW_{\text{NG supply}} + Q \cdot GW_{\text{heat supply}} + P_{\text{el}} \cdot GW_{\text{electricity supply}} + m_{CO_2\text{ emissions}}, \tag{4.9}$$

and

$$FD_{\text{fossil-based}} = m_{\text{NG}} \cdot FD_{\text{NG supply}} + Q \cdot FD_{\text{heat supply}} + P_{\text{el}} \cdot FD_{\text{electricity supply}}. \tag{4.10}$$

3. In the final step, the maximum impact reductions are calculated for global warming ($GW^{\text{max}}_{\text{reduction}}$) and fossil depletion ($FD^{\text{max}}_{\text{reduction}}$) from:

$$GW^{\text{max}}_{\text{reduction}} = GW_{\text{fossil-based}} - GW_{CO_2\text{-based}}, \tag{4.11}$$

and

$$FD^{\text{max}}_{\text{reduction}} = FD_{\text{fossil-based}} - FD_{CO_2\text{-based}}. \tag{4.12}$$

In step 3, the environmental impacts determined in step 1 ($GW_{CO_2\text{-based}}$ and $FD_{CO_2\text{-based}}$) and step 2 ($GW_{\text{fossil-based}}$ and $FD_{\text{fossil-based}}$) are used.

The computed values for $GW^{\text{max}}_{\text{reduction}}$ and $FD^{\text{max}}_{\text{reduction}}$ allow to rank the CO_2-based processes for C1-chemicals according to the environmental impact reduction per kg hydrogen used (see Section 4.4.3).

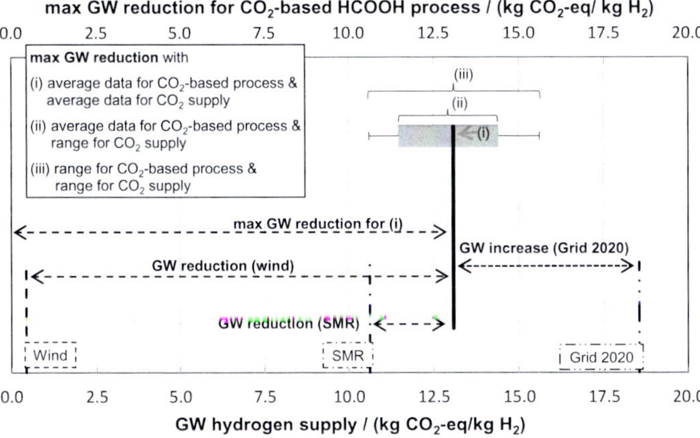

Figure 4.3: Maximum global warming (GW) impact reduction for CO_2-based formic acid processes per kg hydrogen used. The black solid line (i) shows the maximum GW reduction for average data for CO_2 supply and CO_2-based process. The gray box (ii) represents the range for the maximum GW reduction considering the range for the global warming impact of CO_2 supply (-0.42 to -0.99 kg CO_2-eq per kg CO_2) while average data is used for CO_2-based process. The error bars (iii) show the full range for the maximum GW reduction considering the range for the global warming impact of CO_2 supply and the range for the process data.

4.4.2 Interpretation of maximum environmental impact reductions

In this section, the interpretation of the maximum environmental impact reductions is exemplarily presented for formic acid (Figure 4.3).

If average data is assumed for CO_2 supply and the CO_2-based processes, the maximum global warming impact reduction is 13.1 kg CO_2-eq/kg H_2 (see (i) in Figure 4.3). If a best case (-0.99 kg CO_2-eq/kg CO_2) and a worst case (-0.42 kg CO_2-eq/kg CO_2) for CO_2 supply is considered, the maximum global warming impact reduction ranges from 11.5 to 14.4 kg CO_2-eq/kg H_2 (see (ii) in Figure 4.3). If additionally the range for the CO_2-based process concepts is considered, the maximum global warming impact reduction ranges from 10.6 to 15.6 kg CO_2-eq/kg H_2 (see (iii) in Figure 4.3).

The maximum environmental impact reduction can be used to address both goals of this chapter:

I) Comparison of the environmental impact reduction for CO_2-based processes

The maximum environmental impact reductions can be directly used to rank the environmental impact reductions of CO_2-based processes. The actual environmental impact reductions (i.e., including hydrogen supply) are not required if all processes use the same hydrogen source, because all processes use the same amount of hydrogen (functional unit). The comparison of the environmental impact reductions is presented in Section 4.4.3 for all considered CO_2-based processes.

Additionally, the actual environmental impact reduction for each CO_2-based process can be determined by subtracting the environmental impact of hydrogen supply from the maximum environmental impact reduction (see Figure 4.3).

II) Determination of the environmental threshold values for hydrogen supply

The maximum environmental impact reductions can be used as threshold values for hydrogen supply. In other words, if the maximum environmental impact reduction for CO_2-based formic acid is higher than the global warming impact of hydrogen supply, the CO_2-based formic acid production achieves lower global warming impacts than the fossil-based formic acid production.

If the best case for CO_2-based process and CO_2 supply is considered, CO_2-based formic acid production achieves lower global warming impacts than the fossil-based formic acid production for hydrogen supply processes with a global warming impact lower than 15.6 kg CO_2-eq/kg H_2. Thus, the CO_2-based formic acid production is environmentally beneficial even if hydrogen is supplied by steam-methane-reforming (10.6 kg CO_2-eq/kg H_2).

However, if the worst case for CO_2-based process and CO_2 supply is considered, the maximum global warming impact reduction decreases to 10.6 kg CO_2-eq/kg H_2. In this case, the supply of hydrogen by steam-methane-reforming would just be sufficient to achieve the same global warming impacts as fossil-based formic acid production.

If hydrogen is supplied by electrolysis using wind electricity, CO_2-based formic acid always reduces global warming impacts compared to fossil-based processes. If hydrogen is supplied by electrolysis using EU-27 grid mix in 2020, CO_2-based formic acid always increases global warming impacts compared to fossil-based processes.

4.4.3 Comparison of maximum environmental impact reductions

In this section, the maximum impact reductions for global warming and fossil depletion are presented for all considered C1-chemicals: formic acid, carbon monoxide, methanol and methane. For each C1-chemical except methane, maximum impact reductions are presented by comparison to 2 benchmarks. For formic acid and carbon monoxide, the 2 maximum impact reductions are derived by comparison to fossil-based carbon monoxide production with the 2 following H_2/CO ratios: 3 and 4.4. For methanol, the 2 maximum impact reductions are derived by comparison to a low and a high efficient fossil-based process. The low efficient process is related to syngas production by steam-methane-reforming and the high efficient process to syngas production by combined reforming. For methane, the EU-27 natural gas supply mix is considered.

Maximum global warming impact reductions of the considered C1-chemicals range from 2.0 to 15.6 kg CO_2-eq/kg H_2 (Figure 4.4). The highest maximum global warming impact reductions are achieved for CO_2-based formic acid and the lowest for CO_2-based methane. Furthermore, Figure 4.4 reveals for all considered C1-chemicals the hydrogen supply processes that lead to environmentally beneficial CO_2-based processes: All hydrogen supply processes (dashed-dotted vertical lines) that are on the left side of the maximum impact reduction lead to environmentally beneficial CO_2-based processes. If the vertical line of a hydrogen supply process intersects the shown maximum impact reduction range, the CO_2 supply determines whether a CO_2-based process is environmentally beneficial. The CO_2 supply can affect the maximum impact reduction by up to 4.2 kg CO_2-eq/kg H_2 (for methanol).

Based on the forecasted global warming impacts of grid electricity in the EU-27, hydrogen supply by SMR will have lower global warming impacts than hydrogen supply by electrolysis with grid electricity at least until 2050.

Formic acid. The CO_2-based formic acid process achieves the highest maximum global warming impact reductions for all considered CO_2-based processes. The high maximum global warming impact reductions for CO_2-based formic acid are primarily due to the high environmental impact of the fossil-based process.

If carbon monoxide for the fossil-based formic acid production is supplied by SMR (H_2/CO=3), the maximum global warming impact reduction ranges from 10.6 to

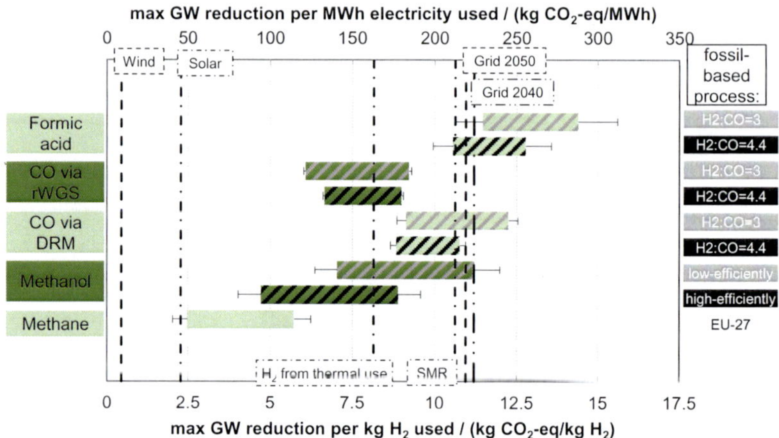

Figure 4.4: Maximum global warming (GW) impact reductions depending on CO_2 supply. The boxes consider a global warming impact of CO_2 supply from -0.42 (left side of box) to -0.99 kg CO_2-eq/kg CO_2 (right side of box). The error bars consider the full range of processes (cf. Figure 4.2). On the bottom axis, the maximum global warming impact reductions are presented per kg H_2. On the top axis, the maximum global warming impact reductions are presented per MWh electricity. In this case, hydrogen is supplied by electrolysis with an energy demand of 50 kWh per kg H_2. The dashed-dotted vertical lines show hydrogen supply processes.

15.6 kg CO_2-eq/kg H_2. Thus, even if hydrogen is supplied by SMR, all considered CO_2-based formic acid processes achieve lower global warming impacts than the fossil-based formic acid processes.

If carbon monoxide for the fossil-based formic acid production is supplied by SMR (H_2/CO=4.4), the maximum global warming impact reduction ranges from 10.0 to 13.6 kg CO_2-eq/kg H_2. The range of the maximum global warming impact reduction is smaller than for the fossil-based process with SMR (H_2/CO=3) because more hydrogen is required for the CO_2-based process and thus the influence of CO_2 supply decreases.

Carbon monoxide (rWGS). The maximum global warming impact reductions range from 6.0 to 9.3 kg CO_2-eq/kg H_2 if fossil-based carbon monoxide is supplied by SMR

($H_2/CO=3$). If fossil-based carbon monoxide is supplied by SMR ($H_2/CO=4.4$), the maximum global warming impact reductions range from 6.7 to 9.0 kg CO_2-eq/kg H_2.

In other words, the rWGS process is environmentally beneficial if hydrogen is supplied by electrolysis using wind or solar electricity but it is not environmentally beneficial if hydrogen is supplied by SMR. If hydrogen from thermal use is employed for rWGS, the CO_2 supply determines whether rWGS is environmentally beneficial or not.

Carbon monoxide (DRM). The maximum global warming impact reductions range from 8.7 to 12.5 kg CO_2-eq/kg H_2. For both fossil-based reference processes, the lower bound of the maximum global warming impact reductions is almost equal. However, the upper bound is higher if fossil-based carbon monoxide is produced by SMR ($H_2/CO=3$) than by SMR ($H_2/CO=4.4$).

For both fossil-based scenarios, the DRM process is environmentally beneficial if hydrogen from thermal use is employed. If the DRM process replaces fossil-based carbon monoxide production by SMR ($H_2/CO=3$) and uses a CO_2 supply process with GW < -0.75 kg CO_2-eq/kg CO_2, the DRM process can even be environmentally beneficial if hydrogen is supplied by SMR.

Depending on the heat supply for the reactor in the DRM process, the maximum global warming impact reductions can vary by about 0.6 kg CO_2-eq/ kg H_2 (error bars). The highest maximum global warming impact reduction is achieved if the reactor is heated with natural gas instead of electricity. The influence of the reactor heating is larger than for the rWGS process because the DRM reaction is more endothermic.

Methanol. For CO_2-based methanol, the maximum global warming impact reductions range from 4.0 to 12.0 kg CO_2-eq/kg H_2. The broad range is caused by both the CO_2-based and the fossil-based processes: On the one hand, the influence of global warming impact of CO_2 supply is highest among all CO_2-based processes, because methanol requires most CO_2 per FU. On the other hand, the range for the efficiency of fossil-based processes is also very large. If a fossil-based process with low efficiency is replaced and a CO_2 source is used with high CO_2 concentration, the CO_2-based methanol production can even be environmentally beneficial if hydrogen from steam-methane-reforming is used.

Methane. CO_2-based methane has the lowest maximum global warming impact reductions of all considered C1-chemicals. Here, only hydrogen from electrolysis using wind electricity always leads to environmentally beneficial CO_2-based processes in all considered scenarios. The supply of hydrogen by electrolysis using solar electricity is not sufficient if CO_2 is supplied by air capture. The CO_2-based methane process is the only process which cannot be environmentally beneficial even if hydrogen from thermal use is employed.

In the Appendix, the maximum global warming impact reductions are also shown for national natural gas supply mixes with lowest and highest global warming impacts in the EU-27 (Figure C.4). A change of the natural gas supply affects the maximum global warming impact reductions of all considered C1-chemicals, because all fossil-based processes use natural gas as feedstock. If the natural gas supply mix with lowest global warming impacts (The Netherlands) is applied, the maximum global warming impact reductions decrease between 0.8–1.2 kg CO_2-eq/kg H_2 for all C1-chemicals. The maximum global warming impact reductions decrease, because mainly the fossil-based processes benefit from the natural gas supply with low global warming impacts. If the natural gas supply mix with highest global warming impacts (Portugal) is applied, the maximum global warming impact reductions increase between 0.7–1.1 kg CO_2-eq/kg H_2 for all C1-chemicals.

Maximum fossil depletion impact reductions of the considered C1-chemicals range from 0.7 to 6.2 kg oil-eq/kg H_2 (Figure 4.5). The results correspond very well to the maximum global warming impact reductions. The order of the maximum fossil depletion impact reductions is exactly the same as for global warming.

The main difference between global warming and fossil depletion impact reductions is that all CO_2-based processes except methane have the potential to achieve lower impacts than fossil-based processes even if hydrogen is supplied by electrolysis with grid mix in 2040 and 2050.

In the Appendix the maximum fossil depletion impact reductions are also shown for the Dutch and the Portuguese natural gas supply mix (Figure C.5). If the Dutch natural gas supply mix is applied, the maximum fossil depletion impact reductions decrease between 0.18–0.29 kg oil-eq/kg H_2 for all C1-chemicals. If the Portuguese natural gas supply mix is applied, the maximum fossil depletion impact reductions increase between 0.15–0.23 kg oil-eq/kg H_2 for all C1-chemicals.

4.5 Conclusions

In this chapter, CO_2-based and fossil-based process concepts are analyzed based on life cycle assessment for the following C1-chemicals: formic acid, carbon monoxide, methanol and methane. For the analysis, the maximum impact reductions for global warming and fossil depletion are derived for the CO_2-based processes. The maximum

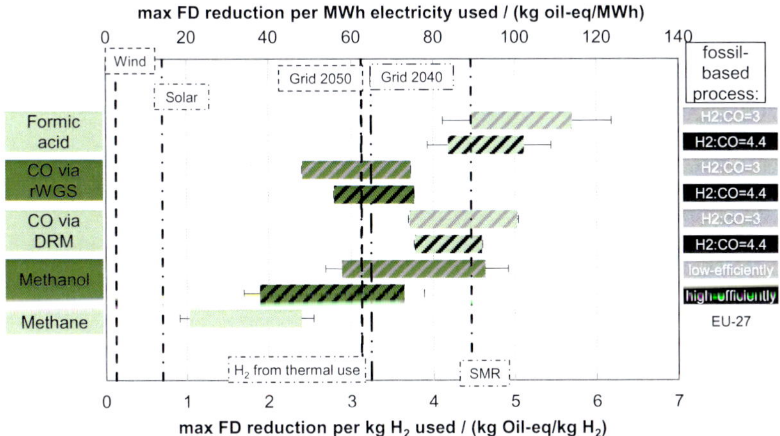

Figure 4.5: Maximum fossil depletion (FD) impact reductions depending on CO_2 supply. The boxes consider a global warming impact of CO_2 supply from 0.24 (left side of box) to 0 kg oil-eq/kg CO_2 (right side of box). The error bars consider the full range of processes (cf. Figure 4.2). On the bottom axis, the maximum fossil depletion impact reductions are presented per kg H_2. On the top axis, the maximum fossil depletion impact reductions are presented per MWh electricity. In this case, hydrogen is supplied by electrolysis with an energy demand of 50 kWh per kg H_2. The dashed-dotted vertical lines show hydrogen supply processes.

environmental impact reduction is the difference of the environmental impact of the fossil-based processes and the CO_2-based processes without considering the environmental impact of hydrogen supply ("free" hydrogen). The maximum environmental impact reduction is presented per kg hydrogen supplied to the CO_2-based processes to identify the CO_2-based process making best of hydrogen.

The comparison of maximum environmental impact reductions for CO_2-based processes is crucial, since most CO_2-based processes are only environmentally beneficial if hydrogen from renewable electricity is used, which is currently limited. From an environmental point of view, hydrogen supplied by additionally installed renewable electricity should be employed for C1-chemicals in the order of highest environmental impact reductions.

The results show that CO_2-based production of formic acid achieves highest environmental impact reductions followed by carbon monoxide and methanol. Lowest environmental impact reductions are achieved for the CO_2-based methane production. Thus, formic acid should be preferably produced from CO_2 if hydrogen supply is limited, e.g., utilization of surplus electricity for hydrogen production.

Furthermore, the maximum environmental impact reduction per kg hydrogen corresponds to the environmental threshold values for hydrogen supply. The CO_2-based processes are environmentally beneficial compared to fossil-based processes if the environmental impact of hydrogen supply is lower than the maximum environmental impact reduction.

If hydrogen is supplied by conventional steam-methane-reforming, only the CO_2-based production of formic acid, carbon monoxide via DRM and methanol have the potential to achieve lower impacts for global warming and fossil depletion compared to the corresponding fossil-based processes. Whether the hydrogen supply by steam-methane-reforming for the CO_2-based processes is actually environmentally beneficial depends on the CO_2 supply and the substituted fossil-based process: The CO_2-based production of formic acid is environmentally beneficial compared to all considered fossil-based processes if CO_2 is captured from flue gases; for DRM and methanol, both very efficient CO_2 supply (flue gas with high CO_2 concentration) and a low-efficient fossil-based process are required. If hydrogen is supplied by electrolysis with renewable electricity (e.g., wind or solar electricity), almost all CO_2-based processes have lower impacts of global warming and fossil depletion than the fossil-based processes. The only exception is CO_2-based methane, which can have higher global warming impacts than natural gas if solar electricity is used and the CO_2 is supplied with high global warming impacts, e.g., air capture using fossil fuels.

For the considered C1-chemicals, the potential to achieve environmentally beneficial CO_2-based processes increases with the number of reaction steps for the fossil-based processes. The CO_2-based production of formic acid seems most promising because the fossil-based process requires several reaction steps and a complex product purification. The CO_2-based production of methane seems most challenging, because the fossil-based counterpart can be directly recovered from underground deposits without additional reaction steps.

Since the CO_2-based production of carbon monoxide, methanol and methane requires hydrogen supply by renewable electricity, the next chapter analysis the hydrogen supply by electrolysis is more detail.

Chapter 5

Power-to-Gas – Use of renewable electricity for CO_2-based processes

The results of the previous chapter show that the environmental impact of hydrogen supply is crucial for all CO_2-based process routes. Only the CO_2-based production of formic acid can achieve lower environmental impacts than the fossil-based process if conventional hydrogen supply by steam-methane-reforming is used. For the CO_2-based production of carbon monoxide, methanol and methane, hydrogen supply by renewable electricity is required to achieve environmentally beneficial processes. Thus, this chapter provides a detailed LCA case study for CO_2-based production of carbon monoxide and methane with hydrogen supply by electrolysis (Power-to-Gas). According to the previous chapter, carbon monoxide has the highest environmental impact reduction among the CO_2-based process which require renewable electricity to be environmentally beneficial and methane the lowest environmental impact reduction. The major goal of this chapter is to determine maximum environmental impact reduction per MWh electricity supply.

In Section 5.1, the LCA method and the considered data are presented for the assessment of the Power-to-Gas routes. In Section 5.2, the results are presented. Finally, conclusions are drawn in Section 5.3.

Major parts of this chapter are reproduced with permission of the American Chemical Society from:

Sternberg, A. and Bardow, A. (2016). Life Cycle Assessment of Power-to-Gas: Syngas vs Methane, *ACS Sustainable Chemistry and Engineering*, 4(8):4156–4165.

5.1 LCA method and data

For the LCA of the Power-to-Gas routes, goal and scope are described in Section 5.1.1. In Section 5.1.2, the data is presented for the technologies considered.

5.1.1 Goal and scope definition

The goal of this chapter is to identify suitable electricity supply processes to achieve environmentally beneficial Power-to-Gas pathways compared to corresponding fossil-based processes. For this purpose, the maximum environmental impact reductions are determined per MWh electricity supply (cf. Section 4.4.1). The considered Power-to-Gas pathways include 2 Power-to-Syngas processes (rWGS and DRM) and 1 Power-to-SNG process (Figure 5.1).

System boundaries and functional unit. For the determination of the maximum environmental impact reductions, the Power-to-Gas pathways are compared to the fossil-based production of syngas and SNG.

The overall functional unit (FU) of the Power-to-Gas pathways is:

- The utilization of 1 MWh electricity for electrolysis.

Both considered Power-to-Syngas processes produce syngas with a molar H_2:CO ratio of 3:1. Conventionally, syngas is produced by steam-methane-reforming (Chen et al., 2005). Thus, both Power-to-Syngas processes are compared to steam-methane-reforming (SMR). In this chapter, the separation of syngas into pure hydrogen and CO is consider as required for subsequent chemical utilization, e.g., in polyurethane production (von der Assen et al., 2015). This combined supply of the two pure products hydrogen and CO in a molar ratio of 3:1 was already considered in Chapter 4. In this chapter, the combined supply of hydrogen and CO is denoted syngas.

The Power-to-SNG process produces a gas containing mostly methane ($>$ 95 vol%) which is fed into the natural gas grid, so-called synthetic natural gas (SNG). The SNG is compared to conventional natural gas supply. Both products SNG and natural gas are considered equivalent based on the lower heating value.

In the comparison of Power-to-Gas pathways to conventional processes, the downstream processes are identical. Therefore, the use-phase and the end-of-life phase of syngas and SNG can be neglected for the present scope. The so-called cradle-to-gate approach is applied considering only the upstream processes for syngas and SNG.

The Power-to-Gas pathways consist of the chemical conversion process, an electrolysis unit and a power plant (Figure 5.1). The power plant supplies the CO_2 for the

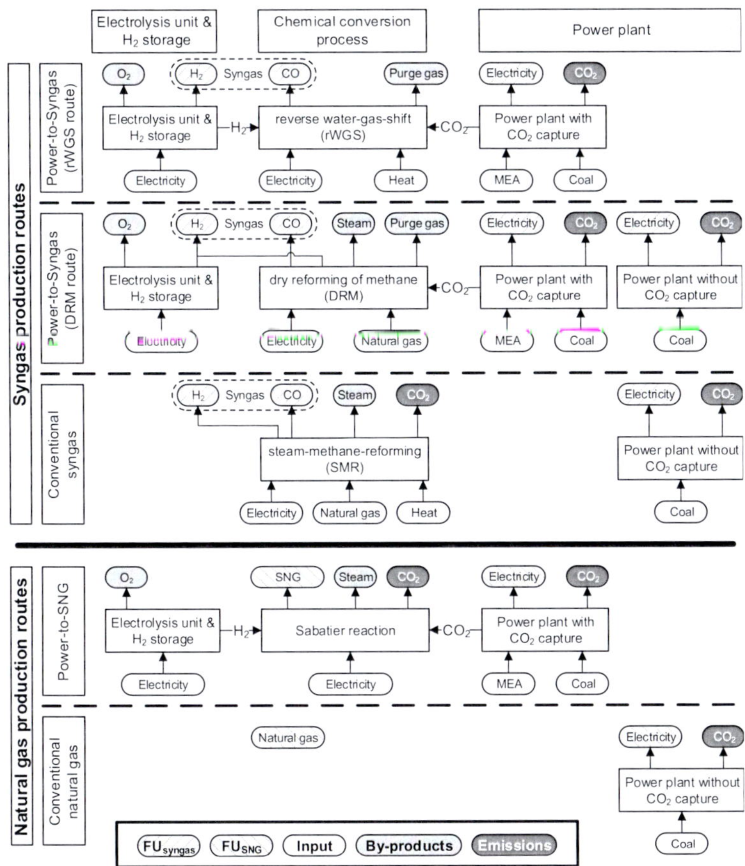

Figure 5.1: Flow chart of considered Power-to-Syngas routes (rWGS and DRM), conventional syngas production, Power-to-SNG and conventional natural gas production. The inputs and outputs (by-products) are used to determine the environmental impact of the processes. The environmental impact of natural gas is directly taken from LCA-database.

chemical conversion process. Power plants are frequently discussed as CO_2 sources because they can supply large amounts of CO_2 (von der Assen et al., 2016). All considered processes are explained in detail in the next section.

The considered Power-to-Gas pathways produce not only the desired products SNG and syngas but also electricity, oxygen, steam and purge gas. A sound comparison requires that all compared processes produce the same products (Jung et al., 2013). Therefore, an identical set of products has to be defined for all compared processes (System expansion, cf. Section 3.3).

For the syngas processes, syngas and electricity are chosen because syngas is the main product of the chemical conversion process, and electricity the main product of the power plant. The amount of electricity results from the amount of CO_2 supplied by the power plant with CO_2 capture. In this case, system expansion avoids the potential misassignment of all environmental benefits to the chemical conversion process only (von der Assen et al., 2013).

Consequently, SNG and electricity are chosen for the SNG process. Here, SNG is considered equivalent to natural gas based on the lower heating value.

The other products (oxygen, steam and purge gas) are defined as by-products according to the approach presented by Jung et al. (2013). The handling of the by-products is explained in the next section. The conventional production processes consist of the chemical conversion process and a power plant (Figure 5.1). Both conventional production processes for SNG and Syngas are extended by a coal-fired power plant without CO_2 capture to produce the electricity required for the selected set of products.

For the comparison of the Power-to-Gas pathways to the corresponding fossil-based processes, the functional unit of using 1 MWh electricity for electrolysis can be converted to the following equivalent functional units that contain the considered set of products for syngas and SNG processes:

- FU_{rWGS}: Supply of 84.3 kg syngas and 93.7 kWh electricity,
- FU_{DRM}: Supply of 146.7 kg syngas and 93.7 kWh electricity,
- FU_{SNG}: Supply of 2.02 GJ SNG and 99.2 kWh electricity.

The functional unit FU_{rWGS} is used for the comparative breakdown analysis of all syngas processes in Section 5.2.1 and is denoted FU_{syngas} in the following.

Environmental indicators. In this chapter, the processes are compared based on their global warming (GW) and fossil depletion (FD) impacts. These impact categories are chosen because they reflect the major motivation for the installation of renewable energies. Further environmental impacts are shown in the Appendix D. The environmental impacts are determined according to ReCiPe 1.08 Midpoint (Hierarchist) (Goedkoop et al., 2009).

5.1.2 Process data for Power-to-Gas

In this section, the considered technologies are briefly presented.

Chemical plant. The CO_2-based and fossil-based production of methane and syngas is based on process data presented in Section 4.2. Since the set of products for carbon monoxide in Chapter 4 also included pure hydrogen, the product corresponds to syngas. Thus, the data can be applied in this chapter. In this thesis, continuous operation of the chemical plants is assumed.

Hydrogen supply. Hydrogen is produced in a proton exchange membrane (PEM) electrolysis unit with an electricity demand of 50 kWh per kg H_2 (DOE, 2014). The electrolysis is operated at 30 bar. The main advantage of PEM electrolysis is a very dynamic operation (Götz et al., 2016; Schiebahn et al., 2015) which enables following intermittent electricity supply, e.g., wind electricity. In this chapter, both (I) steady-state and (II) part-load operation are consider for electrolysis. Steady-state operation requires continuous electricity supply, e.g., through the utilization of grid electricity. Part-load operation is required if intermittent renewable electricity is directly fed into the electrolysis unit. For example, full load hours for solar and wind electricity in Germany are between 900–1,100 and 1,400–5,000 hours per year, respectively (Kaltschmitt et al., 2013). For part-load operation, 2,500 full load hours are considered in this chapter. The influence of full load hours on results is analyzed in the Appendix D.

Furthermore, part-load operation of electrolysis requires hydrogen storage. In this case, storage of hydrogen is assumed in pressure tanks at 25 bar. In this chapter, the hydrogen storage is sized to cover 10 days without intermittent renewable electricity. The influence of the size of the hydrogen storage is investigated in the Appendix D.

Besides hydrogen, electrolysis produces also pure oxygen (8 kg O_2 / kg H_2).

CO_2 supply (power plant). The CO_2 required for the Power-to-Gas pathways is supplied by a coal-fired power plant with CO_2 capture. The efficiency of the considered power plant is 27.3 % (based on the lower heating value) and 90 % of the CO_2 emissions are captured (DoE case 10) (Gerdes et al., 2013). The production of monoethanolamine (MEA) as CO_2 capture solvent is also considered. Most of the MEA is recycled but about 0.0019 kg MEA / kg CO_2 captured is lost (Schreiber et al., 2009).

The electricity in the functional units FU_{syngas} and FU_{SNG} is exactly the electricity produced in a power plant with CO_2 capture to provide the required amount of CO_2 for the rWGS process and SNG process, respectively.

For the conventional processes without CO_2 demand, electricity is produced in a coal-fired power plant without CO_2 capture. The efficiency of the power plant without CO_2 capture based on the lower heating value is 38.3 % (DoE case 9) (Gerdes et al., 2013).

The DRM process requires only about half of the CO_2 of the rWGS process. To still supply the same electricity in the functional unit, the DRM process uses both the power plant with CO_2 capture and the power plant without CO_2 capture.

Utilities. The heat required for the chemical conversion process is supplied by steam from natural gas with an efficiency of 90 % (GaBi ts, 2016). For the electricity supply, most results are presented as function of the environmental impacts. However, for some results, environmental impacts of grid electricity are required. In this case, forecasted EU-27 grid electricity mixes are applied for 2020 and 2030 from the GaBi database (GaBi ts, 2016). The considered LCA data sets for the utilities are summarized in Table B.1.

Construction. For the Power-to-Gas pathways, the environmental impacts for construction are considered for the units of electrolysis and hydrogen storage. The construction of chemical plants and power plants is neglected due to a lack of data. For fossil-based steady-state processes, the impact of construction is usually low compared to the impacts during operation (ecoinvent, 2007a). For the chemical plants of the Power-to-Gas processes, we also consider a steady-state operation. Furthermore, both Power-to-Gas and conventional processes employ similar chemical plants and power plants. Thus, for a comparative assessment, these impacts are of minor importance.

For the construction of the electrolysis unit, LCA data is used based on a PEM fuel cell. For the construction of the hydrogen storage pressure tanks, only the steel demand is considered following Mori et al. (2014). Details for the construction of electrolysis and hydrogen storage are summarized in Table D.1.

By-products. In this chapter, an environmental credit for by-products is not considered, because utilization cannot always be achieved. The influence of environmental credits is investigated in Appendix D.

5.2 Results and discussion

In this section, the impacts for global warming and fossil depletion of Power-to-Gas pathways are compared to the corresponding conventional processes. Based on the comparison, maximum environmental impact reductions per MWh electricity supply are derived. First, results are presented for steady-state operation of electrolysis and then for part-load operation.

5.2.1 Steady-state operation of electrolysis

Power-to-Syngas vs conventional syngas production. Figure 5.2 shows the impacts for global warming and fossil depletion of Power-to-Syngas (rWGS and DRM) and conventional syngas production for the production of 84.3 kg syngas and 93.7 kWh electricity (FU_{syngas}). Here, the forecasted EU-27 grid electricity mix for 2020 is applied for electricity supply.

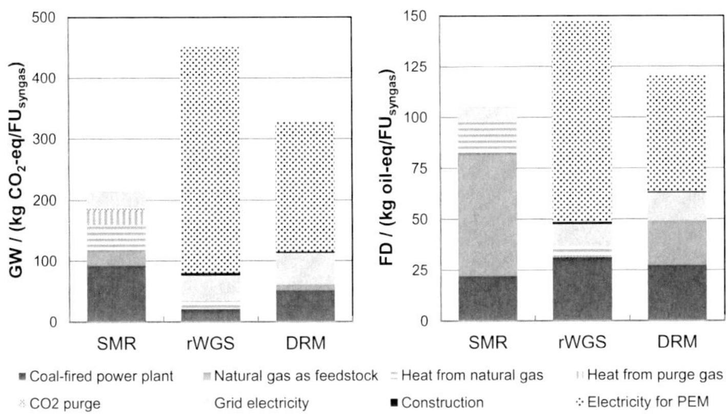

Figure 5.2: Global warming (GW) impact (left) and fossil depletion (FD) impact (right) of Power-to-Syngas (rWGS and DRM) and conventional syngas production (SMR). The functional unit (FU_{syngas}) is the production of 84.3 kg syngas and 93.7 kWh electricity. The bars of the single processes show the breakdown of the impacts. The environmental impacts for grid electricity are based on forecasted EU-27 electricity mixes in 2020.

The global warming impact of conventional syngas production is 215 kg CO_2-eq/FU_{syngas}. The functional unit FU_{syngas} includes electricity and syngas. About 43 % of the global warming impact is due to the electricity production in the coal-fired power plant power and 57 % due to the syngas production. The global warming impact of the syngas production stems from electricity supply (13 %), heat supply from natural gas (19 %) and purge gas (11 %), CO_2 purge (2 %) and supply of natural gas as feedstock (12 %).

For the rWGS process, the global warming impact is about 451 kg CO_2-eq/FU_{syngas} in 2020. About 82 % of the global warming impact is due to the electricity supply for electrolysis. Hence, the rWGS process will strongly benefit from a future reduction of global warming impacts for electricity supply. The remaining 18 % of the global warming impact is mainly due to the electricity supply for the chemical conversion process, emissions of the coal-fired power plant and the heat supply. The CO_2 emissions in the power plant are lower than in conventional syngas production because CO_2 is captured and used in the chemical conversion process. The global warming impact of the construction of the electrolysis unit is negligible if the process is operated continuously (< 1 %).

For the DRM process, the global warming impact is about 328 kg CO_2-eq/FU_{syngas} in 2020. The global warming impact is lower than for the rWGS process, because the DRM process requires less hydrogen. Thus, less electricity is required for the electrolysis. About 65 % of the global warming impact of the DRM process is due to electricity supply for electrolysis. The remaining 35 % of the global warming impact of the DRM process are mainly due to electricity supply for the chemical conversion process, emissions of the power plant and the natural gas supply. The global warming impact of the power plant is higher compared to the rWGS process, because only half of the CO_2 is captured from the power plant compared to the rWGS process (see above).

The fossil depletion impact of the conventional syngas production is about 105 kg oil-eq/FU_{syngas}. About 20 % of the fossil resources are required for electricity production and 80 % for syngas production. For the rWGS and DRM processes, the fossil depletion impacts are 147 and 121 kg oil-eq/FU_{syngas}, respectively.

Thus, both Power-to-Syngas processes are expected to lead to higher GHG emissions and higher fossil resource depletion than conventional syngas production in 2020. This is mainly due to the impact of electricity supply. Since the environmental impacts from electricity supply should be reduced in the future, the impact from electricity supply on Power-to-Syngas is studied in detail.

Figure 5.3 shows the global warming impact of conventional syngas production and both Power-to-Syngas processes as function of the global warming impact of electricity supply. For both Power-to-Syngas processes, the global warming impact of electricity supply has a strong influence on the total global warming impact, because both processes require large amounts of electricity for electrolysis. For the DRM process, the influence of the global warming impact of electricity supply is slightly lower than for

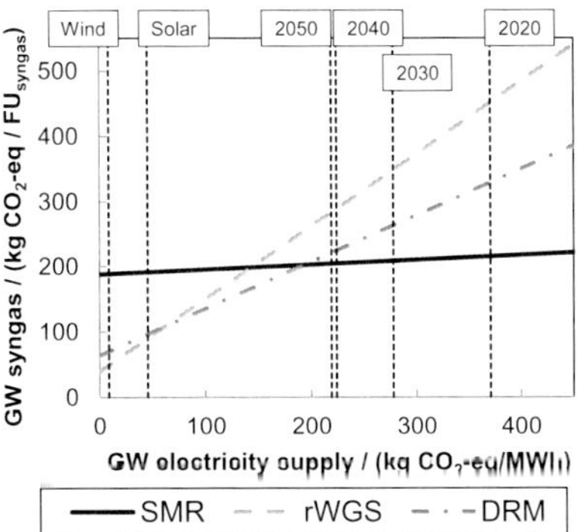

Figure 5.3: Global warming impact of conventional syngas (SMR) and Power-to-Syngas (rWGS and DRM) as function of global warming impact of electricity supply. The vertical lines represent the global warming impacts of electricity supply from wind and solar, and forecasted EU-27 electricity mixes.

the rWGS process, because the DRM process requires less hydrogen. For electricity supply with reduced GHG emissions, Power-to-Syngas becomes beneficial over conventional production. However, based on the forecasted EU-27 electricity mixes, this would be achieved until 2050.

For conventional syngas production, the total global warming impact depends much less on the global warming impact of electricity supply, because the conventional syngas process requires electricity only for the compression and separation of gases.

The intersections of the rWGS and DRM process with the conventional syngas production are the threshold values for global warming impact of electricity supply. As explained in Section 3.5, this threshold value corresponds to the maximum global warming impact reductions per MWh electricity supply. The maximum global warming impact reductions are 143 and 194 kg CO_2-eq/MWh for rWGS and DRM process, respectively.

The maximum fossil depletion impact reductions can be determined in the same way as for global warming. The maximum fossil depletion impact reductions are 58 and 75 kg oil-eq/MWh for rWGS and DRM process, respectively.

In the Appendix D, the dependence of the maximum impact reductions for global warming and fossil depletion (Figure D.1) is investigated on utilization of by-products,

CO_2 supply and efficiency of electrolysis. For both Power-to-Syngas processes, the maximum global warming impact reductions decrease by 70 kg CO_2-eq/MWh, if CO_2 supply does not avoid CO_2 emissions. CO_2 emissions are not avoided if, for example, CO_2 is used which otherwise would be stored underground. CO_2 emissions are also not avoided if the installation of a power plant with CO_2 capture delays installation of renewable electricity.

The analysis of 12 further environmental impact categories shows that environmental impact reductions are only achieved for photochemical oxidant formation (rWGS and DRM). For Power-to-Syngas processes, the environmental hot spots of most impact categories are construction of electrolysis unit and coal supply. However, uncertainty is also higher in most other impact categories compared to global warming and fossil depletion impacts (Hauschild et al., 2013).

Power-to-SNG vs conventional natural gas production. Figure 5.4 shows the global warming and the fossil depletion impact of Power-to-SNG and conventional natural gas production for the production of 2.02 GJ SNG and 99.2 kWh electricity (FU_{SNG}) in the year 2020.

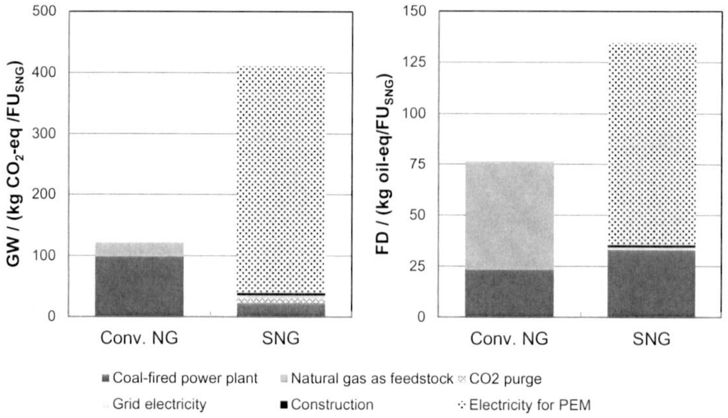

Figure 5.4: Global warming (GW) impact (right) and fossil depletion (FD) impact (left) of Power-to-SNG and conventional natural gas (conv. NG) production. The functional unit (FU_{SNG}) is the production of 2.02 GJ SNG and 99.2 kWh electricity. The bars of the single processes show the breakdown of the impacts. The environmental impacts for grid electricity are based on forecasted EU-27 electricity mixes in 2020.

The global warming impact of conventional natural gas production is 121 kg CO_2-eq/FU_{SNG}. 81 % of the emissions occur during electricity production in the coal-fired power plant and 19 % during natural gas supply. For Power-to-SNG, the global warming impact is about 410 kg CO_2-eq/FU_{SNG} in 2020. Only about 10 % of the global warming impact is due to emissions of the power plant, CO_2 purge and electricity supply. About 90 % of the global warming impact is due to the electricity supply for electrolysis. The maximum global warming impact reduction is 86 kg CO_2-eq/MWh (Figure 5.5).

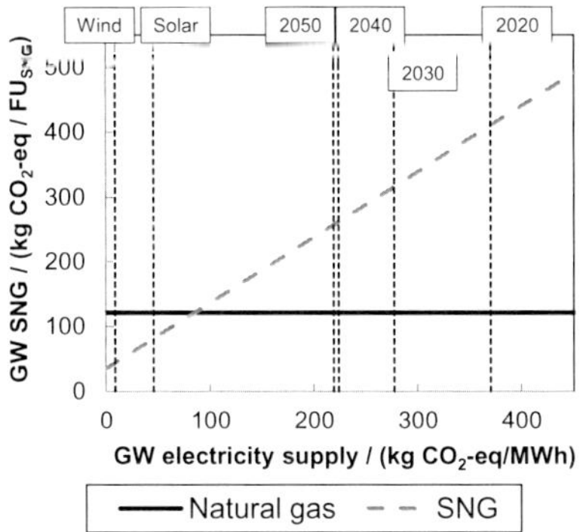

Figure 5.5: Global warming impact of conventional natural gas and Power-to-SNG as function of global warming impact of electricity supply. The vertical lines represent the global warming impacts of electricity supply from wind and solar, and forecasted EU-27 electricity mixes.

The fossil depletion impact of the conventional natural gas production is 76 kg oil-eq/FU_{SNG}. 30 % of the fossil depletion impacts are due to the coal-fired power plant and 70 % due to the natural gas supply. For Power-to-SNG, the fossil depletion impact is 134 kg oil-eq/FU_{SNG}. The maximum fossil depletion impact reduction is 42 kg oil-eq/MWh.

For Power-to-SNG, the sensitivity of environmental impact reductions is presented depending on utilization of by-products, CO_2 supply and efficiency of electrolysis in the Appendix B. The results are similar to Power-to-Syngas. Power-to-SNG benefits most from an environmental credit for by-products, because a large amount of heat is produced during the exothermic reaction. In contrast to Power-to-Syngas, the maximum global warming impact reduction becomes even negative if CO_2 is used which otherwise would be stored. In this case, Power-to-SNG would thus have higher global warming impacts than conventional natural gas supply even if global warming impacts for electricity supply were zero.

5.2.2 Part-load operation of electrolysis

In this section, the influence of part-load operation of electrolysis is investigated. Part-load operation of electrolysis enables the direct utilization of intermittent renewable electricity. Through the direct utilization of renewable electricity such as wind or solar, a reduction of global warming and fossil depletion impacts can be achieved already today. However, if Power-to-Gas pathways use exclusively renewable electricity, the electrolysis unit has only low full load hours and huge hydrogen storage capacities are required. Through the combination of intermittent renewable and base-load grid electricity, the full load hours of electrolysis can be increased and the size of the hydrogen storage can be decreased. In the following, intermittent renewable electricity is denoted renewable electricity and base-load grid electricity only grid electricity. The combination of renewable ($EI_{\text{renewable electricity}}$) and grid electricity ($EI_{\text{grid electricity}}$) is environmentally beneficial if the sum of their environmental impacts is lower than the maximum environmental impact reductions per MWh electricity supply ($EI^{\text{max}}_{\text{reduction}}$) :

$$EI^{\text{max}}_{\text{reduction}} \geq EI_{\text{renewable electricity}} \cdot X + EI_{\text{grid electricity}} \cdot (1 - X) \qquad (5.1)$$

Based on inequality (5.1), the maximum environmental impact reductions can be determined per MWh *renewable* electricity ($EI^{\text{max}}_{\text{reduction, renewable}}$) or per MWh grid electricity ($EI^{\text{max}}_{\text{reduction, grid}}$). In Figure 5.6, the maximum impact reductions for global warming and fossil depletion are shown per MWh renewable electricity as function of the share (X) of renewable electricity used for electrolysis. For the environmental impact of grid electricity, two scenarios based on forecasted EU-27 grid electricity mixes are considered: 2020 and 2030.

Global warming impacts. For all Power-to-Gas pathways, the maximum global warming impact reduction per MWh renewable electricity is largest if the electrolysis uses 100 % renewable electricity. In 2020, the DRM process has the highest maximum global warming impact reduction with 143 kg CO_2-eq/MWh, followed by the rWGS and SNG process with 105 and 52 kg CO_2-eq/MWh, respectively. The maximum global warming impact reduction for 100 % renewable electricity differs from the maximum global warming impact reduction determined in the previous section due to lower full load hours for the electrolysis unit, the additionally required hydrogen storage and the utilization of grid electricity for the chemical conversion process. For

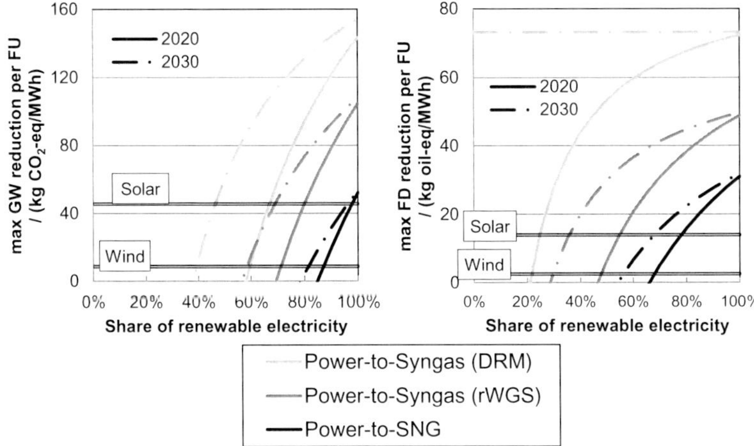

Figure 5.6: Maximum reduction of global warming (GW) (left) and fossil depletion (FD) impact (right) per MWh renewable electricity as function of the share of renewable electricity used for electrolysis. The solid lines represent the scenario 2020 for the grid electricity. The dashed lines represent a scenario for 2030. In 2030, the DRM process does not require renewable electricity to achieve lower fossil depletion impacts than the fossil-based process. Thus, the maximum environmental impact reduction does not depend on the share of renewable electricity used for electrolysis The vertical axis can also be used to present electricity supply processes (cf. Figure 4.3). In this case, the horizontal lines represent the environmental impacts of electricity supply from the renewable sources wind and solar.

the utilization of 100 % renewable electricity, environmental impacts of wind and solar electricity are lower than the maximum global warming impact reductions for all Power-to-Gas pathways. In other words, all Power-to-Gas pathways reduce global warming impacts compared to fossil-based processes if wind or solar electricity is employed. Among the considered Power-to-Gas pathways, the DRM process achieves the highest global warming impact reduction per MWh renewable electricity for all shares of renewable electricity and for all considered years. The global warming impact reduction is the difference of the maximum global warming impact reduction and the environmental impact of renewable electricity. For example, if wind electricity is used for electrolysis in 2020, the global warming impact reduction for DRM process is

130 kg CO_2-eq per MWh renewable electricity.

Figure 5.6 also shows the minimum shares of renewable electricity that are required to be environmentally beneficial depending on the renewable electricity technology used for electrolysis. If wind electricity is used for electrolysis in 2020, the minimum required shares of wind electricity for DRM, rWGS, and SNG process in 2020 are 60 %, 70 % and 88 %, respectively.

In 2030, the maximum global warming impact reductions for 100 % renewable electricity are almost the same as for 2020. If the electrolysis unit uses 100 % renewable electricity, the different global warming impacts for the electricity supply in 2020 and 2030 only affect the electricity supply required for the chemical plants. The electricity demand of the chemical plant is low compared to the electricity demand of the electrolysis unit. However, the minimum required share of renewable electricity to be environmentally beneficial decreases compared to 2020.

Fossil depletion impacts. The results for the fossil depletion impact correspond qualitatively well to the global warming impact. However, quantitatively, much lower shares of renewable electricity are required compared to global warming impacts to be environmentally beneficial with a lowest value of 20 % for the DRM process in 2020. The maximum fossil depletion impact reductions for 100 % renewable electricity for DRM, rWGS and SNG in 2020 are 65, 49 and 33 kg oil-eq/MWh, respectively. Based on the forecasted EU-27 electricity mixes, no renewable electricity is required for DRM in 2030 to be environmentally beneficial.

In the Appendix D, a sensitivity analysis is presented for utilization of by-products, CO_2 supply, efficiency of electrolysis, and construction of electrolysis unit and hydrogen storage. Regarding the construction of electrolysis and hydrogen storage, the size of the hydrogen storage has the highest influence on maximum impact reductions for global warming and fossil depletion (Figure D.2). However, the results for construction should only be considered as indicative because currently only very few LCA data sets are available for the construction of hydrogen storage and electrolysis unit.

5.3 Conclusions

In this chapter, the following three Power-to-Gas pathways are analyzed: Power-to-Syngas via rWGS and via DRM as well as Power-to-SNG. Syngas and SNG are synthesized from conversion of CO_2 and hydrogen. As the CO_2 source, a coal-fired power plant with CO_2 capture is considered. Hydrogen is supplied by water electrolysis. For operation of the electrolyzer, both steady-state and part-load operation are considered. The Power-to-Gas pathways are compared based on the maximum environmental impact reductions per MWh electricity.

For steady-state operation of electrolysis, the maximum global warming impact reductions are 194, 143 and 86 kg CO_2-eq/MWh for DRM, rWGS and SNG process, respectively. Based on the forecasted EU-27 electricity mixes, no Power-to-Gas route achieves lower global warming impacts than the corresponding conventional processes until 2050 if only grid electricity is used for electrolysis.

Through part-load operation of electrolysis, intermittent renewable electricity can be directly used in electrolysis. If the electrolysis unit uses only renewable electricity, the maximum global warming impact reductions are 143, 104 and 52 kg CO_2-eq/MWh for DRM, rWGS and SNG process, respectively. Thus, all Power-to-Gas pathways have lower global warming impacts compared to corresponding conventional processes if 100 % wind (12 kg CO_2-eq/MWh) or solar (46 kg CO_2-eq/MWh) electricity is used for electrolysis. If all Power-to-Gas pathways use the same amount of wind electricity, the DRM process achieves the highest environmental impact reduction. Since the exclusive utilization of renewable electricity leads to low full load hours for electrolysis and requires huge storage capacities, a hybrid strategy combining grid electricity and renewable electricity is also investigated. If a mix of wind electricity and EU-27 grid mix in 2020 is used for electrolysis, DRM, rWGS and SNG process require at least 60 %, 70 % and 88 % of wind electricity to reduce global warming impacts compared to the corresponding conventional process.

Besides electricity supply, the CO_2 supply is also crucial for the environmental performance of Power-to-Gas pathways. Power-to-Gas pathways can achieve lower global warming impacts than conventional processes if CO_2 from a coal-fired power plant is captured which would emit the CO_2 otherwise. However, the global warming impact reduction of the Power-to-Gas pathways decreases if the utilized CO_2 does not avoid CO_2 emissions (e.g., if CO_2 storage is avoided). In this case, the SNG process has even higher global warming impacts than conventional natural gas supply.

Overall, both considered Power-to-Gas pathways can decrease global warming impacts and fossil depletion impacts if the majority of electricity is supplied by renewable electricity. However, currently the demand of renewable electricity exceeds the supply. Thus, a comparison to alternative options for the utilization of renewable electricity is required. This comparison is presented in the next chapter.

Chapter 6

Power-to-X – Assessment of utilization options for renewable electricity

In Chapter 5, it is shown that the Power-to-Chemical routes do not achieve lower global warming impacts than the fossil-based counterparts if grid electricity is used until 2050. Environmentally beneficial Power-to-Chemical routes can be achieved if renewable power is used. However, renewable power will be limited even in future. Therefore, this chapter compares and ranks several technologies that can use renewable power (Power-to-X.) In short, this chapter tries to answer the question: Given 1 MWh of renewable electricity which Power-to-X system brings the greatest environmental benefit? Finally, the environmental assessment is combined with an economic assessment and CO_2 mitigation costs are determined.

The proposed approach for the comparison of Power-to-X systems is presented in Section 6.1. In Section 6.2, a systematic overview of the considered Power-to-X systems is provided and the results are presented. Finally, conclusions for the environmental assessment of Power-to-X systems are drawn in Section 6.3.

Major parts of this chapter are reproduced with permission from The Royal Society of Chemistry from:

Sternberg, A. and Bardow, A. (2015). Power-to-What? – Environmental Assessment of Energy Storage systems, *Energy & Environmental Science*, 8(2):389-400.

6.1 LCA method for comparative assessment of Power-to-X systems

In this chapter, the concept of maximum reduction of environmental impacts is applied for Power-to-X systems with different products (Figure 6.1). The use of renewable power for Power-to-X systems is considered in the EU-27 (average data of all members) and additionally in a variety of countries including the United States, Brazil, Germany, Great Britain, China and Japan. The *maximum reduction of environmental impacts* for the utilization of *renewable power* presented in this chapter is identical to the *reduction of environmental impacts* for the utilization of *surplus power* as presented in Sternberg and Bardow (2015).

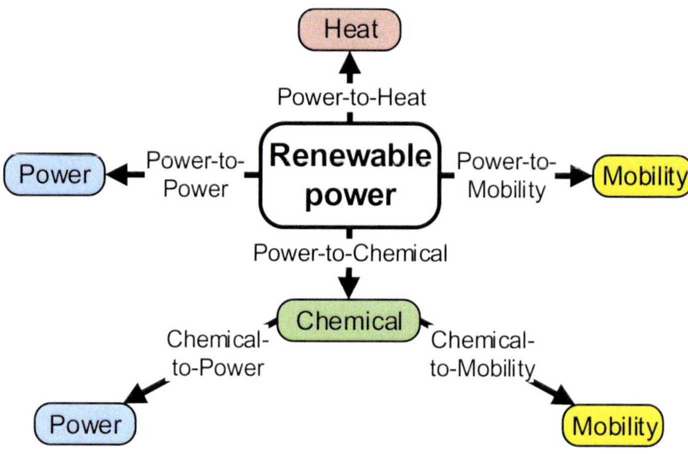

Figure 6.1: Overview of considered Power-to-X systems categorized by their products

In this chapter, the functional unit of the Power-to-X systems is the use of 1 MWh renewable electricity. The Power-to-X systems are compared in detail by the following environmental impact categories: global warming (GW) and fossil depletion (FD). These impacts are chosen because they are the key drivers for renewable energy systems. The environmental impacts are determined according to ReCiPe 1.08 Midpoint (Hierarchist) (Goedkoop et al., 2009) because ReCiPe provides an indicator for the depletion of fossil fuels and global warming. The results for fossil depletion and global warming impact are presented in oil-equivalents and CO_2-equivalents, respectively. Results for the impact categories eutrophication (freshwater and marine), human toxicity, ionizing radiation, mineral resource depletion, photochemical oxidant formation, ozone depletion, particulate matter and terrestrial acidification have also been analyzed and are given in the Appendix. Water footprint is also an important potential impact and should be considered once assessment methods are sufficiently well established (Gu et al., 2014) and sufficient data for Power-to-X systems are available.

Most accurate results are achieved if site-specific conditions are considered for both the Power-to-X systems (e.g. heat and power supply) and the substituted products (Henriksson et al., 2014). However, LCA-databases typically provide country-specific values (ecoinvent Data V 2.2, 2010; GaBi ts, 2016). For this reason, a country-specific approach is used to minimize the effort of data acquisition.

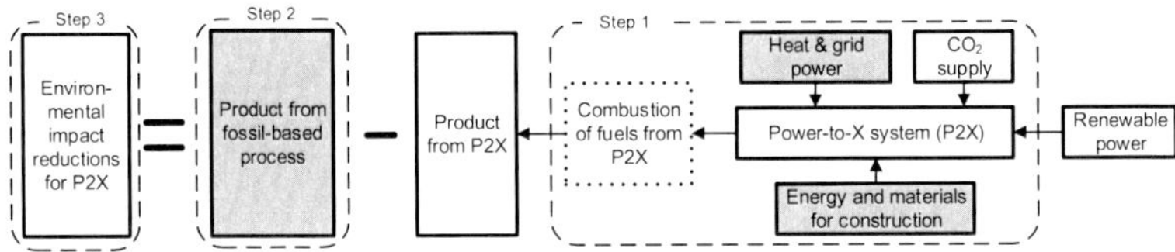

Figure 6.2: The maximum environmental impact reductions for Power-to-X systems (P2X) are the difference between the environmental impacts of products from fossil-based processes and from Power-to-X systems. The environmental impacts of Power-to-X systems consider the processes: (i) heat and grid power supply, (ii) CO_2 supply, (iii) construction, and (iv) combustion of fuels. The combustion of fuels from Power-to-Fuel systems must be considered if the fuel is converted to power or used for mobility. The environmental impacts of processes in the gray boxes are modeled using aggregated data from LCA databases. The amount of product gained from the Power-to-X system is defined by the use of 1 MWh renewable electricity.

Determination of the maximum reduction of environmental impacts for Power-to-X systems. In this section, the maximum reduction of environmental impacts (cf. Section 4.4.1) is adopted to Power-to-X systems (Figure 6.2):

1. The Power-to-X systems are modeled to determine input and output flows. All Power-to-X systems use 1 MWh of renewable electricity. The input and output

flows are then used to determine the global warming impact (GW_{P2X}) and fossil depletion impact (FD_{P2X}) of the Power-to-X systems (P2X) according to

$$GW_{\text{P2X}} = Q \cdot GW_{\text{heat supply}} + P_{\text{grid}} \cdot GW_{\text{grid power supply}} + m_{\text{CO}_2} \cdot GW_{\text{CO}_2\text{ supply}} + P_{\text{capacity}} \cdot GW_{\text{construction}} + M_{\text{product}} \cdot GW_{\text{combustion}}, \quad (6.1)$$

and

$$FD_{\text{P2X}} = Q \cdot FD_{\text{heat supply}} + P_{\text{grid}} \cdot FD_{\text{grid power supply}} + m_{\text{CO}_2} \cdot FD_{\text{CO}_2\text{ supply}} + P_{\text{capacity}} \cdot FD_{\text{construction}}. \quad (6.2)$$

Here, the input flows are heat (Q), grid power (P_{grid}) and CO_2 (m_{CO_2}). The environmental impacts (GW_{supply} and FD_{supply}) for the supply process are explained in Section 6.2.2. The environmental impacts from construction ($GW_{\text{construction}}$ and $FD_{\text{construction}}$) are based on the capacity (P_{capacity}) of the Power-to-X systems. The output flow is the product (M_{product}) of each Power-to-X systems. Emissions from combustion ($GW_{\text{combustion}}$) of products are taken into account for utilization of CO_2-based chemicals as energy carriers for electricity generation (Chemical-to-Power) and transportation purposes (Chemical-to-Mobility). The environmental impacts of construction and combustion are also explained in Section 6.2.2.

2. For each product of the Power-to-X system, the replaced fossil-based process is identified. The environmental impacts of the replaced fossil-based processes are determined from LCA-databases (see section 5.3.3).

3. In the final step, the maximum global warming impact reduction ($GW_{\text{reduction}}$) and maximum fossil depletion impact reduction ($FD_{\text{reduction}}$) are calculated from:

$$GW^{\text{max}}_{\text{reduction}} = M_{\text{product}} \cdot GW_{\text{fossil-based process}} - GW_{\text{P2X}}, \quad (6.3)$$

and

$$FD^{\text{max}}_{\text{reduction}} = M_{\text{product}} \cdot FD_{\text{fossil-based process}} - FD_{\text{P2X}}. \quad (6.4)$$

Here, the amount of product (M_{product}) and environmental impacts of Power-to-X system (GW_{P2X} and FD_{P2X}) determined in the first step are used. The environmental impacts of the replaced fossil-based processes ($GW_{\text{fossil-based process}}$ and $FD_{\text{fossil-based process}}$) are determined in step 2.

The computed values for $GW^{\text{max}}_{\text{reduction}}$ and $FD^{\text{max}}_{\text{reduction}}$ allow to rank the Power-to-X systems according to the environmental impact reduction (see Sec. 5.3.4).

6.2 Comparison of environmental impact reductions for Power-to-X systems

In this section, a case study for selected Power-to-X systems is presented using LCA data sets for different countries. All considered countries are presented in the Appendix (see Tables B.6–B.9). In Section 6.2.1, a systematic overview and a short description including current efficiencies are provided for the Power-to-X systems considered. All Power-to-X systems include the opportunity to store renewable electricity. The storage options are also briefly discussed. In Sections 6.2.2–6.2.4, the 3 steps of the proposed method for a comparative environmental assessment are applied to the case study. In Section 6.2.5, the CO_2 mitigation costs for the EU-27 are presented.

6.2.1 Considered Power-to-X systems

Power-to-Power. Power-to-Power systems convert electricity to chemical or mechanical energy, which is stored and later reconverted to electric power. Today, large scale Power-to-Power is mostly based on pumped hydro storage (PHS). PHS achieves efficiencies of 65–80 % but it is limited by geographical constraints (Ibrahim et al., 2008; Díaz-González et al., 2012; Ferreira et al., 2013).

The other Power-to-Power technology already operating at large-scale is compressed air energy storage (CAES). However, currently only 2 plants are in operation worldwide (Chatzivasileiadi et al., 2013). These plants require 0.7–0.8 kW h input electricity and 1.2–1.3 kW h natural gas to generate 1 kW h electricity (Ibrahim et al., 2008; Denholm and Kulcinski, 2004). The natural gas demand for CAES can be avoided by storing the heat released during air compression (Foley and Lobera, 2013): the stored heat is then used to preheat the air before it is expanded in a gas turbine. These so-called adiabatic compressed air energy storage systems are supposed to achieve efficiencies of about 75 % (Ferreira et al., 2013) but they are still under development. Therefore, non-adiabatic CAES are considered. CAES is also limited geographically by the availability of suitable salt caverns (Beaudin et al., 2010).

Batteries are a mature energy storage technology for small-scale applications (Díaz-González et al., 2012; Ibrahim et al., 2008; Ferreira et al., 2013). A promising technology for medium scale storage are flow batteries, storing energy in solutions. Flow batteries are characterized by high efficiencies of 65–85 %, very low self-discharge and high cycle capability (Ibrahim et al., 2008; Díaz-González et al., 2012; Ferreira et al., 2013). A drawback of flow batteries is the low volumetric energy density compared to other batteries.

Power-to-Mobility. Power-to-Mobility systems store electricity in battery electric vehicles (BEV). The most discussed storage units for battery electric vehicles are lithium-ion and nickel metal-hydride batteries (Majeau-Bettez et al., 2011). In this chapter, a lithium-ion battery is considered. This technology seems to be the most promising based on its high efficiency of about 80 % and its high cycling capability (Ibrahim et al., 2008; Díaz-González et al., 2012; Ferreira et al., 2013). Battery electric vehicles consume 0.14–0.2 kWh/km (Metz and Doetsch, 2012; Majeau-Bettez et al., 2011). Although first commercial vehicles are already available, the technology is still in development. Hennings et al. (2013) investigated the utilization of surplus wind power in electric vehicles.

Power-to-Heat. Power-to-Heat systems convert power directly to heat. The ratio of the heat obtained to the power input is the characteristic efficiency of these systems, the coefficient of performance (COP). Mature technologies to generate heat are electric boilers and heat pumps. The maximum COP for an electric boiler is 1. A more efficient Power-to-Heat technology are heat pumps. Braungardt et al. (2013) report annual COPs of air source heat pumps of 2.1–3.4, and of ground source heat pumps of 2.6–4.9 based on field tests. The generated heat can be stored directly in buildings or in hot water tanks. Hedegaard et al. (2012) and Hewitt (2012) investigated the potential of heat pumps to integrate wind power.

Power-to-Chemical. Power-to-Chemical systems first produce hydrogen by electrolysis. A suitable technology to absorb the fluctuating renewable power from renewable energies is proton exchange membrane (PEM) electrolysis. 1 MWh electricity allows the production of 18–22 kg hydrogen (DOE, 2014). Hydrogen can be used directly as fuel, as chemical feedstock or converted to other fuels.

If hydrogen is used as fuel or chemical feedstock, the fluctuating production has usually to be compensated by storage. Hydrogen storage requires huge storage volumes because of the low volumetric energy density of hydrogen. The construction of additional hydrogen storage capacities can be avoided by directly feeding hydrogen into the already existing natural gas grid (feed-in of H_2). However, the fraction of hydrogen in the natural gas grid is limited, for example in Germany to 5–12 vol% (Müller et al., 2012).

Conversion of hydrogen with CO_2 to C1-chemcials. An alternative option to deal with the low energy content of hydrogen is the conversion to C1-chemicals with higher energy content. In this chapter, the conversion to methane, methanol and syngas are considered. To satisfy the annual global demand of hydrogen and methanol about 3,200–5,400 TWh and 575–950 TWh are required, respectively. For syngas, no data on the global demand is available. For syngas use in polyurethane production only, a demand of 25–40 TWh would be required (see Table E.2 in Appendix). In this chapter, the conversion of chemicals is also considered to 2 further products:

Chemical-to-Power. Reconversion of fuels to power in fuel cells is considered. Fuel cells achieve an efficiency of 45–60 % (Peighambardoust et al., 2010).

Chemical-to-Mobility. The utilization of fuel for vehicles is also investigated. Hydrogen can be used in a fuel cell vehicle with a consumption of about 0.34–0.36 kWh per km (Granovskii et al., 2006; Offer et al., 2011). Methane and methanol can be used in a vehicle with an internal combustion engine. These engines consume about 0.52 kWh per km (Ou et al., 2010) and 0.44 kWh per km (Rose et al., 2013), respectively.

6.2.2 Environmental impacts of Power-to-X systems (step 1)

To determine the environmental impacts, the Power-to-X systems described in Section 6.2.1 are modeled in (GaBi ts, 2016). The models are based on efficiencies. For each technology, the range of efficiencies presented in Section 6.2.1 is considered. Hydrogen conversion is modeled using literature data obtained through process simulations. For the hydrogen conversion processes, an uncertainty of 10 % is assumed for the heat and grid power demand. Detailed information about the used data can be found in the Appendix E.2. The processes considered for the determination of the environmental impacts of the Power-to-X system are shown in Figure 6.2 and explained in the following:

(i) *Heat and grid power supply.* Heat is only required for syngas production and compressed air energy storage. Syngas production requires heat to drive the endothermic reverse water-gas-shift reaction. Compressed air energy storage requires heat to run gas turbines. In both cases, it is assumed that heat is provided by natural gas combustion. An additional grid power demand occurs for the hydrogen conversion processes to compress the gases and to separate the products. For both heat from natural gas and grid power, LCA data sets are considered for several countries (see Tables B.6 & B.7 in Appendix).

(ii) *CO_2 supply.* The conversion of hydrogen to C1-chemicals requires CO_2 as carbon source. 2 scenarios are considered for the CO_2 supply:

- CO_2 emissions avoided: In this scenario, CO_2 is used which otherwise would be emitted to the atmosphere. The environmental impact of the CO_2 supply depends on the CO_2 capture process. 2 cases are considered:
  - Ideal case: the CO_2 is captured from a process where CO_2 is already separated with high purity but then emitted to the atmosphere. Hence, no

additional energy is required to capture the CO_2. In this case, the global warming impact of CO_2 supply is 1 kg CO_2-eq per kg CO_2 and all other environmental impacts of CO_2 supply are zero.

– Realistic case: the CO_2 is captured from a coal fired power plant with a net power output of 550 MW. Based on the DoE case 10 (Gerdes et al., 2013), the efficiency (based on lower heating value) of the power plant is 27.3 %, and 90 % of the CO_2 is captured. The environmental impacts of the power plant with CO_2 capture cannot be attributed only to CO_2 supply because the power plant provides 2 products: power and CO_2 (von der Assen et al., 2013, 2014). To determine the environmental impact of CO_2 supply, a credit is given for avoiding environmental impacts of power generation in a power plant without CO_2 capture. The coal-fired power plant without CO_2 capture has a net efficiency (based on lower heating value) of 38.3 % and the total CO_2 is emitted to the atmosphere (DoE case 9, (Gerdes et al., 2013)). The environmental impacts of both coal supply and coal combustion are calculated country-specific (see Table B.8 in Appendix). For the EU-27, the global warming and fossil depletion impact of CO_2 supply is −0.67 kg CO_2-eq per kg CO_2 and 0.08 kg oil-eq per kg CO_2, respectively.

• CO_2 storage avoided: In this scenario, CO_2 is used which otherwise would be stored. Hence, no additional energy is required to capture the CO_2 but also no CO_2 emissions are avoided. Thus, the environmental impact of CO_2 supply is zero.

(iii) *Construction of plants and storage units.* The environmental impacts of construction are considered for storage units, electrolysis, fuel cells and hydrogen conversion processes. For fossil-based industrial steady-state processes using fossil fuels, the environmental impacts of construction are usually of minor importance compared to the overall environmental life cycle impacts (ecoinvent, 2007a). In this chapter, however, construction is considered for the Power-to-X systems because of the fluctuating operation. This fluctuating operation requires larger plants due to lower full load hours compared to operation in steady-state. Construction is not considered for vehicles using methane and methanol, as well as for vehicles using diesel or gasoline. For battery electric vehicles, only the construction of the lithium-ion battery is considered. Furthermore, it is assumed that the specific environmental impacts for construction of plants and storage units are not affected by the size of the plant and storage unit. The employed values are based on literature data and database ecoinvent Data V 2.2 (2010). The values are reported in detail in the Appendix (Table E.5).

(iv) *Combustion of fuel.* For Power-to-Fuel systems, the emissions of combustion have to be taken into account if the fuel is combusted, i.e., converted to power or used for mobility. CO_2 emissions are calculated by assuming complete combustion.

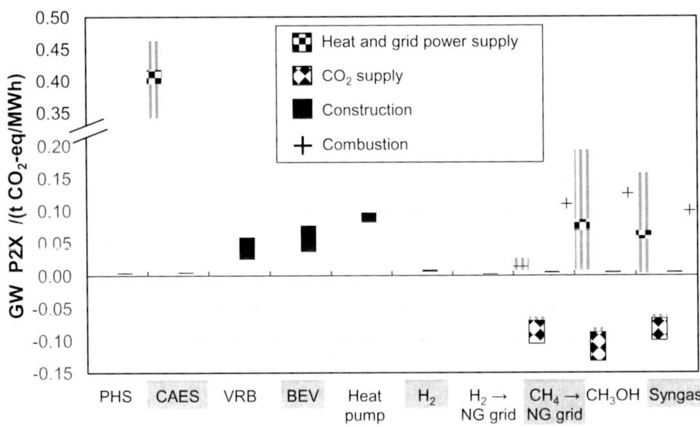

Figure 6.3: Breakdown of global warming (GW) impact of Power-to-X systems into considered processes. All values are per MWh electricity input. The narrow ranges (patterned) represent LCA data sets for the EU-27 considering the technology uncertainties (see Table E.3–E.5). The broad ranges (gray striped) further cover the full range of LCA data sets for all countries considered. If only 1 range is shown, global warming impacts are independent from country-specific LCA data sets. For the global warming impact of fuel combustion, a single value is shown. For CO_2 supply, only the global warming impact for the scenario CO_2 emissions avoided is shown. For the scenario CO_2 storage avoided, the corresponding global warming impact would be zero.

Global warming impact of Power-to-X systems. The global warming impact of Power-to-X systems are computed per MWh electricity input according to Equation (6.1). Figure 6.3 shows the breakdown of the global warming impact of Power-to-X systems into the considered processes: heat and grid power supply, CO_2 supply, construction and combustion of fuel (cf. Figure 6.2). The global warming impacts span a range due to both the considered technology uncertainties (e.g., range of efficiencies) and the use of LCA data sets for different countries. These 2 sources of variation are visualized as follows: the narrow ranges (patterned) represent average LCA data sets for the EU-27 considering the technology uncertainties. The broad ranges (gray striped) further cover the full range of LCA data sets for all countries. If only 1 range is shown, global warming impacts are independent from country-specific LCA data sets. The global warming impact of combustion is a fixed value because CO_2 emissions for complete combustion per kg feedstock are independent from efficiencies and country-specific LCA data sets.

For pumped hydro storage, global warming impact is only due to construction (about 0.005 t CO_2-eq per MWh). Compressed air energy storage (CAES) has nearly the same global warming impact during construction as pumped hydro storage. Additionally 0.40–0.42 t CO_2-eq per MWh arise from heat supply for the EU-27. This value drops to 0.34 t CO_2-eq per MWh for Norway which is the country with the lowest global warming impact for heat from natural gas. CAES have the highest global warming impact for heat and power supply of all Power-to-X systems. However, the additional heat supply allows CAES to be the only considered Power-to-X system which produces more than 1 MWh electricity. Batteries as well as heat pumps induce global warming impacts only during their construction. Here, the heat pump has the highest global warming impact (0.08–0.1 t CO_2-eq per MWh) followed by the lithium-ion battery and the vanadium redox flow battery. For the heat pump, the majority of global warming impacts are caused by the supply of the refrigerant (R134a) and the emission of the refrigerant at the end-of-life. For the direct utilization and the feed-in of H_2, global warming impact is also only related to construction. For the conversion of hydrogen to C1-chemicals, global warming impact occurs during construction, and from the supply of heat and grid power. For all C1-chemicals, the global warming impact during construction is in the same order of magnitude as for compressed air energy storage. Regarding heat and grid power supply, methanol and syngas have the highest global warming impact (0.06–0.09 t CO_2-eq per MWh for the EU-27) among the C1-chemicals. The global warming impact for heat and grid power supply strongly depends on the considered country. Grid power has the highest impact. All hydrogen conversion processes get a credit for the utilization of CO_2 in the scenario CO_2 emissions avoided. Methanol production receives the largest credit (up to 0.13 t CO_2-eq per MWh) because it uses most CO_2 per converted hydrogen. In the scenario CO_2 storage avoided, the global warming impact of CO_2 supply is zero. If C1-chemicals are converted to power or used for mobility, the global warming impact during combustion has to be taken into account. The highest global warming impact arises for the combustion of methanol (0.13 t CO_2-eq per MWh).

Fossil depletion impact of Power-to-X systems. The contributions from construction, heat and grid power supply, and CO_2 supply (scenario: CO_2 storage avoided) to fossil depletion impact correspond well with global warming impact. In contrast to global warming impact, the fossil depletion impact for CO_2 supply (scenario: CO_2 emissions avoided) is positive because the power plant with CO_2 capture has a lower efficiency than the power plant without CO_2 capture (see Figure E.1 in Appendix).

6.2.3 Identification of fossil-based processes replaced by Power-to-X systems (step 2)

In this section, fossil-based processes are identified which are replaced by the Power-to-X systems. For each product, the fossil-based process with the highest marginal costs is replaced assuming market behavior according to the industry cost curve (Spitz, 2003). Processes with co-products are not considered because of the remaining co-product demand. The LCA data sets used for the fossil-based processes can be found in the Appendix (Table B.4).

Power-to-Power systems replaces power from a gas turbine with an efficiency of 40 %. This process is chosen because it is a peak-load technology with high costs and it is in this regard comparable to the power generation from Power-to-Power systems (e.g. PHS) (Sensfuß et al., 2008).

Power-to-Mobility systems replace gasoline and diesel vehicles. Both types of fuel are considered because the diffusion of gasoline and diesel strongly depends on the considered country. The fuel consumption of the replaced gasoline and diesel vehicles is 0.52 kWh/km and 0.38 kWh/km, respectively (GaBi ts, 2016).

For Power-to-Heat systems, it is assumed that heat generation by natural gas boilers with an efficiency of 91 % is replaced. In 2020, natural gas boilers will have the highest costs of fossil-based heating systems (Nitsch et al., 2010).

Power-to-Hydrogen systems replaces hydrogen from natural gas steam reforming. This process is the main industrial process with only hydrogen as product (Baufumé et al., 2013).

For the feed-in of H_2 or CH_4, 2 cases are considered: in the first case, the natural gas consumption mix does not change due to the feed-in of H_2 or CH_4. Hence, the utilization of the natural gas is irrelevant and the replaced fossil-based process is fossil natural gas supply. For the feed-in of H_2, it is also considered that the combustion of hydrogen emits no greenhouse gases in contrast to the combustion of natural gas. In the second case, it is assumed that the feed-in of H_2 or CH_4 leads to an increased natural gas demand in the mobility sector. The automotive industry could promote the utilization of feed-in H_2 or CH_4 to reduce the CO_2 emissions of their fleets. In this case, the replaced fossil-based process is a gasoline or diesel vehicle. Power-to-CH_3OH systems replaces the natural-gas-based production of methanol because this is the typical industrial scale process (Jadhav et al., 2014).

Syngas from the conversion of hydrogen and CO_2 replaces syngas from natural gas steam reforming because this is the industrial scale standard process (Chen et al., 2005).

In the Appendix, the replaced technologies are varied and the resulting environmental impacts are shown for Power-to-Heat (Figure E.11), Power-to-Mobility (Figure E.12) and Power-to-Power (Figure E.13).

6.2.4 Global warming and fossil depletion impact reductions for Power-to-X systems (step 3)

Global warming impact reductions. The global warming impact reductions are shown for all considered Power-to-X systems in Figure 6.4. The heat pump and the battery electric vehicle have the highest global warming impact reduction potential (up to 1.38 t CO_2-eq per MWh and 1.14 t CO_2-eq per MWh, respectively). For heat pumps, a wide range of COPs is considered because the COP of a heat pump depends on many factors (e.g. type of heat pump, required heating temperature). Heat losses during storage are not considered. A technology with a COP of 1 like an electric boiler would still save about 0.3 t CO_2-eq per MWh (see Figure E.11 in Appendix). The battery electric vehicle saves most greenhouse gases if gasoline vehicles are replaced because gasoline vehicles have a lower efficiency than diesel vehicles. Here, it is assumed that the storage unit of the heat pump and battery electric vehicle is large enough to store the product from using intermittent renewable power. The Power-to-Power systems (PHS, CAES, and VRB) save between 0.28 and 0.59 t CO_2-eq per MWh.

The global warming impact reductions of the Power-to-Fuel systems are much lower than for the other Power-to-X systems: the global warming impact reductions are even less than for Power-to-Heat with a COP of 1.

Among the Power-to-Chemical systems, the direct utilization of hydrogen without any further conversion achieves the highest impact reductions. All considered direct utilization routes for hydrogen lead to very similar global warming impact reductions. The feed-in of H_2 saves 0.13–0.20 t CO_2-eq per MWh if the natural gas consumption mix does not change due to the feed-in of H_2. If the resulting natural gas is used as fuel for mobility, this process will save 0.14–0.30 t CO_2-eq per MWh. The impact

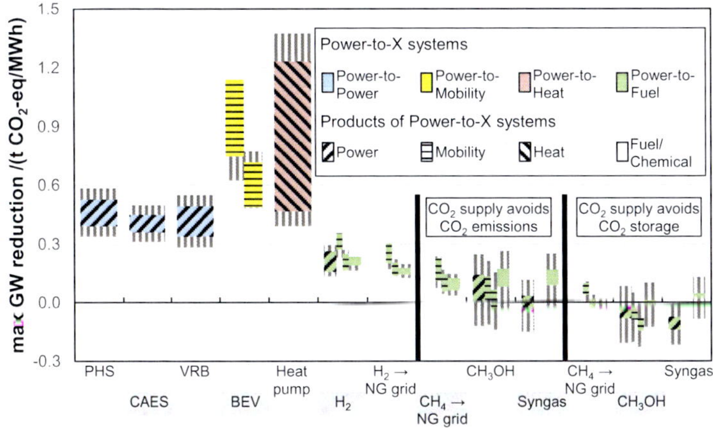

Figure 6.4: Global warming (GW) impact reduction for the considered Power-to-X systems. The narrow ranges (colored) represent LCA data sets for the EU-27 considering the technology uncertainties (see Table E.3–E.5). The broad ranges (gray striped) further include the full range of LCA data sets for all countries considered. For Power-to-Mobility and Fuel-to-Mobility, impact reductions are shown for 2 replaced fossil-based processes: the left bar represents gasoline vehicles and the right bar represents diesel vehicles. For Power-to-Fuel systems, impact reductions are shown for the different utilization routes of the products. For the C1-chemical routes, the global warming impact reductions are shown for both CO_2 supply scenarios (CO_2 emissions avoided and CO_2 storage avoided).

reductions for the feed-in of H_2 into the natural gas grid and the subsequent use as fuel for mobility are lower than for the direct use of hydrogen as fuel for mobility because the feed-in H_2 is converted with a lower efficiency than pure hydrogen.

For the C1-chemical routes, the 2 scenarios for the credit of CO_2 supply have to be distinguished (cf. Section 6.2.2): the C1-chemical routes achieve highest impact reductions if CO_2 emissions are avoided. However, the impact reductions are lower than for the direct use of hydrogen due to additional power demand and conversion losses. The feed-in of CH_4 into the natural gas grid saves 0.03–0.15 t CO_2-eq per MWh. The utilization of fuel for mobility saves 0.04–0.24 t CO_2-eq per MWh.

The results for methanol strongly depend on country-specific LCA data. This is mainly due to the grid power demand. For all utilization routes of methanol, the global warming impact can even increase. Among the utilization routes for methanol, mobility has the lowest global warming impact reductions because the efficiency for conversion of methanol and gasoline are nearly the same. In contrast, for the conversion of methanol to power, a higher efficiency is assumed than for the fossil-based process.

For syngas, the results also strongly depend on country-specific LCA data. The production of syngas and the subsequent utilization as feedstock saves up to 0.25 t CO_2-eq per MWh. The reconversion of syngas to power saves no CO_2 because the production of syngas has a high energy demand for the separation of the products. However, this separation is only required if the syngas is used as feedstock but not if it is reconverted to power. This example demonstrates that an assessment based on reconversion to power is in general not appropriate for C1-chemicals used as chemical feedstock. If CO_2 utilization avoids CO_2 storage, the impact reductions for all C1-chemical routes decrease. For the EU-27, global warming impact reductions are only obtained by the feed-in of CH_4 and the utilization of syngas as chemical feedstock.

Fossil depletion impact reductions. The fossil depletion impact reductions for the Power-to-X systems are shown in Figure 6.5. The results correspond well to the global warming impact reductions (Figure 6.4), except for the Power-to Fuel systems using CO_2 (case: CO_2 storage avoided). However, the gap between Power-to-Heat systems and battery electric vehicles increases. Also, the gap between battery electric vehicles and Power-to-Power systems decreases compared to global warming impact reductions. This is due to the different replaced fossil fuels: all Power-to-Power and Power-to-Heat systems replace natural gas which has lower specific emission (kg CO_2-eq per MJ) than gasoline and diesel.

For the C1-chemical routes, the fossil depletion impact reductions are the same for both scenarios CO_2 emissions avoided (ideal case) and CO_2 storage avoided as no additional energy is required for CO_2 capture. However, the fossil depletion impact reductions are lower for the scenario CO_2 emissions avoided (realistic case) than for the scenario CO_2 storage avoided due to the additional coal required for CO_2 capture.

Further environmental impact reductions. Further environmental impact reductions are presented in the Appendix (Figures E.2–E.10). For these impact categories, the impact of construction is neglected due to lack of data. The variation of the LCA data due to the different countries is larger than for global warming and fossil depletion impact reductions. Environmental impact reductions are always achieved by Power-to-Heat, Power-to- Mobility, Power-to-Power and Power-to-Hydrogen systems. As for global warming and fossil depletion impacts, Power-to-Heat and Power-to-

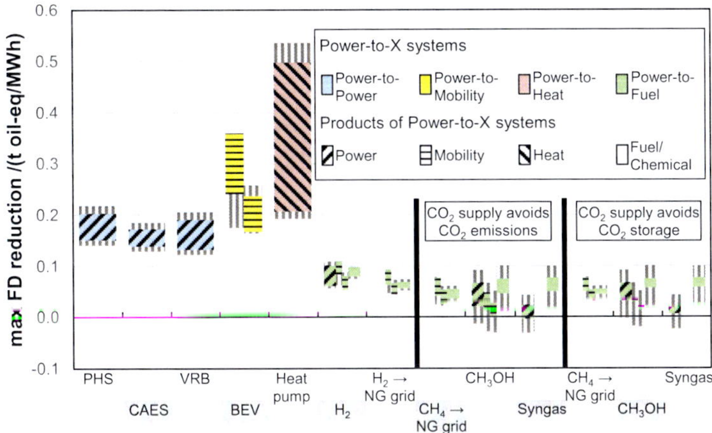

Figure 6.5: Fossil depletion (FD) impact reductions for the considered Power-to-X systems. The narrow ranges (colored) represent LCA data sets for the EU-27 considering the technology uncertainties (see Table E.3–E.5). The broad ranges (gray striped) further include the full range of LCA data sets for all countries considered. For Power-to-Mobility and Fuel-to-Mobility, impact reductions are shown for 2 replaced fossil-based processes: the left bar represents gasoline vehicles and the right bar represents diesel vehicles. For Power-to-Fuel systems, the impact reductions are shown for the different utilization routes of the products. For the C1-chemical routes, the fossil depletion impact reductions are shown for both CO_2 supply scenarios (CO_2 emissions avoided and CO_2 storage avoided).

Mobility systems achieve highest impact reductions. The C1-chemical routes reduce environmental impacts for freshwater eutrophication and mineral resource depletion. For marine eutrophication, human toxicity, ozone depletion, particulate matter and photochemical oxidant formation, reductions are not always achieved and depend on the country. For terrestrial acidification and ionizing radiation, the C1-chemical routes increase the environmental impacts.

6.2.5 CO_2 mitigation costs

In this section, the environmental assessment is complemented with an economic assessment calculating CO_2 mitigation costs. For Power-to-X systems, CO_2 mitigation costs are the ratio of additional costs compared to the fossil-based process producing the same product and the global warming impact reduction determined in Section 6.2.4:

$$\text{CO}_2 \text{ mitigation costs} = \frac{(LPC_{\text{P2X}} - LPC_{\text{conv}}) \cdot M_{\text{product}}}{GW_{\text{reduction}}}. \tag{6.5}$$

Here, the amount of product per MWh renewable electricity (M_{product}) and the global warming impact reduction ($GW_{\text{reduction}}$) are used from step 1 and 3 in section 2. The levelized product costs (LPC) are the minimum prices required to cover all expenses of production, *i.e.*, the ratio of total annual costs and the annual amount of products. The total annual costs consider operating costs and the capital costs multiplied with the capital recovery factor (CRF). An interest rate of 6 % is applied.

CO_2 mitigation costs are calculated for the EU-27 in 2020. Hence, EU-27-specific global warming impact reductions are used. The data used for calculation of the levelized product costs for Power-to-X systems (LPC_{P2X}) and for fossil-based processes (LPC_{conv}) are presented in Tables E.6–E.8 in the Appendix. A constant demand is assumed for all products. The costs for the renewable power supply are zero.

For the Power-to-Chemical systems, capital costs are only considered for hydrogen production. The capital cost for the hydrogen conversion processes (methane, methanol and syngas) are not considered due to a lack of data. However, for the CO_2 mitigation costs only the difference between LPC_{P2X} and LPC_{conv} is required. Hence, capital costs are also neglected for the fossil-based production of methane, methanol and syngas. For the battery electric vehicle, the battery is sized for a daily travel distance of 30 km.

The levelized product costs for the Power-to-X systems strongly depend on 2 factors: when and how often is intermittent renewable power available. Therefore, the CO_2 mitigation costs are presented as function of the available renewable power. Figure 6.6 shows the mean value for the CO_2 mitigation costs if renewable power occurs once a day. The case for renewable power occurring once a week is presented in Figure E.14 in the Appendix. The available renewable power has the highest influence on CO_2 mitigation costs for Power-to-Chemical systems and the heat pump. This implies that the levelized product costs for these Power-to-X systems are dominated by the capital costs for electrolysis and heat pump. The CO_2 mitigation costs for the battery electric vehicle are independent from the available renewable power because the capital costs of the battery only depend on the energy capacity costs.

The lowest CO_2 mitigation costs are found for the battery electric vehicle. The CO_2 mitigation costs are even negative, i.e., the reduction of CO_2 emissions is economically beneficial. The second lowest CO_2 mitigation costs are found for the electric boiler. The electric boiler achieves lower CO_2 mitigation costs than the heat pump because the investment costs are about 5 times lower while the mean COP is only a factor of 3.5. For less hours with renewable power supply, the electric boiler is followed by Power-to-Power systems. Only for more than 6,000 hours with renewable power supply, the heat pump achieves third lowest CO_2 mitigation costs. The highest CO_2 mitigation costs occur for Power-to-Chemical systems. Among the Power-to-Fuel systems, Power-to-H_2 has the lowest CO_2 mitigation costs and production of methane highest CO_2 mitigation costs.

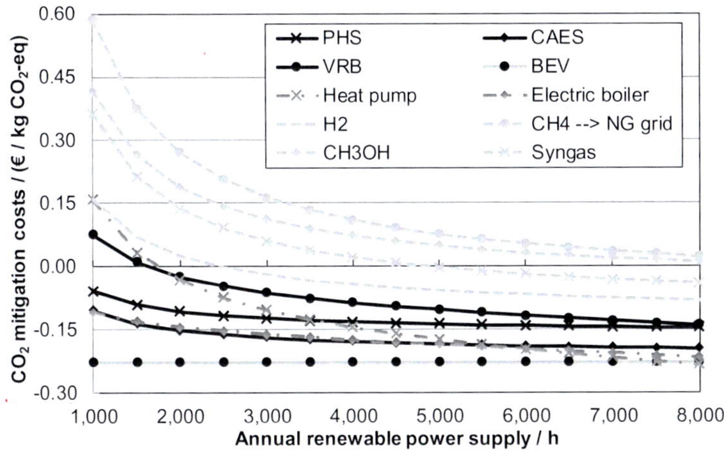

Figure 6.6: CO_2 mitigation costs of Power-to-X systems as a function of available renewable power. In this case, renewable power occurs once a day.

If the renewable power occurs only once a week (see Figure E.14 in the Appendix), CO_2 mitigation costs increase for all Power-to-X systems. The increased period without renewable power supply has the highest influence on vanadium redox flow battery, battery electric vehicle and the systems using heat storage (heat pump and electric boiler). In this case, the compressed air energy storage and the pumped hydro storage achieve the lowest CO_2 mitigation costs.

This cost analysis can only be indicative due to the high uncertainty for the novel technologies considered, the uncertainty is studied for the CO_2 mitigation costs in the Appendix (Figure E.15).

6.3 Conclusions

In this chapter, a general method is presented based on life cycle assessment to compare Power-to-X systems with the products power, heat, mobility, fuel and chemical feedstock. The presented method accounts for the actual use of the products. The results for the considered Power-to-X systems show that the highest global warming and fossil depletion impact reductions are achieved by using intermittent renewable power in heat pumps (Power-to-Heat) and battery electric vehicles (Power-to-Mobility). The third highest environmental impact reductions are achieved by the Power-to-Power systems. Environmental impact reductions are also achieved for the direct utilization of hydrogen. However, these impact reductions are significantly lower than for the Power-to-Heat, Power-to-Mobility and Power-to-Power systems. The climate benefit of converting hydrogen into C1-chemicals depends strongly on the credit of the CO_2 supply: the utilization of CO_2 only contributes to global warming impact reductions if the CO_2 supply avoids emissions. If CO_2 is used which otherwise would be stored, greenhouse gas emissions even increase in most cases. In contrast to greenhouse gas emissions, the C1-chemical routes decrease fossil depletion for both scenarios of CO_2 supply.

An estimate of costs indicates that lowest CO_2 mitigation costs are achieved by compressed air energy storage and pumped hydro storage. These Power-to-X systems are even economically beneficial for less than 1,000 hours renewable power supply and if the renewable power occurs only once a week. The battery electric vehicle and the Power-to-Heat systems also have the potential for low CO_2 mitigation costs but they require a more frequent renewable power supply (e.g., once a day) and more hours with renewable power supply. The presented method allows decision makers to identify environmentally and economically promising Power-to-X system depending on country-specific conditions.

Overall, the results show that the utilization of renewable electricity for CO_2-based production of C1-chemicals is not the most efficient way to use renewable electricity in environmental terms. Furthermore, CO_2-based production of C1-chemicals has the highest CO_2 mitigation costs. Thus, in the next chapter it is investigated if biomass-based processes are more suited for the substitution of fossil-based feedstocks in the chemical industry.

Chapter 7

CO_2 vs Biomass – Comparison of renewable carbon sources

Besides CO_2, biomass is a further renewable carbon feedstock for the production of C1-chemicals. However, currently, the main focus of biomass utilization is on the utilization as fuel for electricity and heat generation, and as automotive fuel. In this chapter, a comparison of CO_2 and biomass as feedstock for chemicals is presented.

In Section 7.1, the challenges are presented for a comparison of CO_2-based and biomass-based processes. In Section 7.2, the optimization-based approach is presented for the comparison of CO_2-based and biomass-based processes. A case study is presented in Section 7.3. Finally, the conclusions are drawn in Section 7.4.

Minor parts of this chapter are reproduced with permission of Elsevier from:

Sternberg, A., Teichgräber, H., Voll, P., and Bardow, A. (2015). CO_2 vs Biomass: Identification of Environmentally Beneficial Processes for Platform Chemicals from Renewable Carbon Sources, *Computer Aided Chemical Engineering*, 37:1361–1366.

7.1 Introduction

To obtain a holistic picture of renewable production of C1-chemicals, CO_2-based processes have to be compared to biomass-based processes. For this purpose, this chapter presents an approach for comparison of CO_2-based and biomass-based processes for C1-chemicals. The challenge for such a comparison is that both CO_2-based and biomass-based processes compete for renewable inputs (i.e., renewable electricity and biomass). Thus, the result of the comparison depends on the alternative utilization of these renewable inputs. Furthermore, the alternative utilization of biomass and renewable electricity strongly depends on regional conditions. The results should thus be presented as a function of these conditions. The comparison of CO_2-based and biomass-based processes is further complicated because the biomass-based processes can also use hydrogen produced from renewable electricity. Thus, biomass-based processes can use 2 renewable inputs for production of C1-chemicals. In this chapter, optimization is used to identify environmentally most beneficial process routes.

7.2 Optimization-based identification of environmentally beneficial process routes

In this chapter, superstructure-based optimization is used to identify environmentally most beneficial process routes. First, a mathematical superstructure model is formulated incorporating all potential process routes (Section 7.2.1). Second, the generated model is analyzed using optimization algorithms to identify process routes minimizing the total environmental impacts (Section 7.2.2).

7.2.1 Superstructure

The superstructure contains the following process routes (cf. Figure 7.1): (I) Biomass-to-Chemicals process routes, (II) CO_2-to-Chemicals process routes, (III) hydrogen production processes, (IV) conventional process routes for chemicals, and (V) alternative utilization of biomass and renewable power. The process routes are modeled by constant conversion factors representing the ratio between inputs and outputs. Note that the process routes are coupled in the superstructure model, enabling an integrated analysis to exploit possible synergies of the considered process routes: Hydrogen is mandatory for the CO_2-to-Chemicals process routes and optional for the Biomass-to-Chemicals process routes; CO_2 for the CO_2-to-Chemicals process routes could also be supplied from biomass-based process routes.

Process route (V) represents alternative utilization of the renewable feedstocks biomass and renewable power outside the chemical industry. An alternative utilization would lead to environmental impact reductions by substituting a conventional process and its related environmental impacts (cf. Chapter 6). For the alternative utilization, no specific technologies are considered. Rather, the results are presented as function of environmental impact reductions for the alternative utilization of biomass and renewable electricity. This enables the identification threshold values for the environmental impact reduction of the alternative utilization pathways. The threshold value indicates when the alternative utilization achieves higher environmental impact reductions than the utilization for production of chemicals.

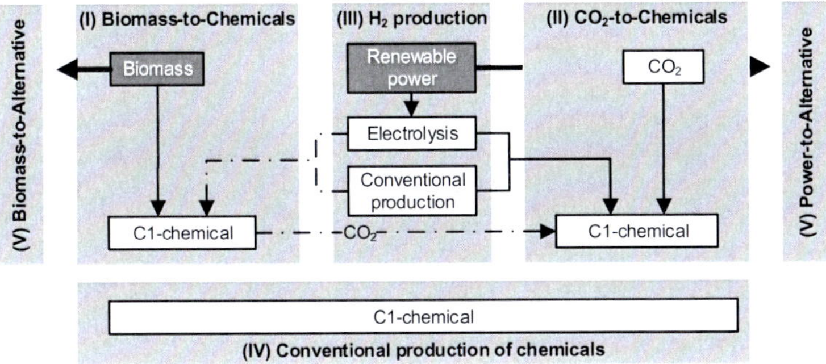

Figure 7.1: Superstructure embedding all considered process routes for C1-chemicals. Dashed/dotted lines indicate optional connections, solid lines indicate mandatory connections.

7.2.2 Linear optimization model

The optimization is carried out minimizing the total environmental impacts of the considered system (cf. Figure 7.1):

$$\min_{x} z = EI^T x, \tag{7.1}$$

$$\text{s.t. } Ax = 0, \tag{7.2}$$

$$x_{\text{RE}} = x_{\text{RE,FU}} \in \mathbb{R}^n \tag{7.3}$$

$$x_{\text{BM}} = x_{\text{BM,FU}} \in \mathbb{R}^n \tag{7.4}$$

The objective function is formulated in terms of the environmental impacts $EI \in \mathbb{R}^n$ of the single processes and the scaling factors $x \in \mathbb{R}^n$ of the single processes (decision variables). The scaling factors x represent the choice and the utilization level (continuous value x) of the single process routes. Note that the scaling factors can also become negative, i.e., a process with a negative scaling factor is substituted. The total environmental impacts are minimized while mass and energy balances are fulfilled (Equation 7.2). Matrix A contains all inputs and outputs of the considered processes including the alternative utilization of biomass and renewable electricity outside the chemical industry. The environmental credits for the alternative utilization of biomass and renewable electricity are considered in the vector EI.

Constraints. The scaling factors for biomass (x_{BM}) and renewable electricity (x_{RE}) supply represent the functional unit (FU) of the system investigated. Thus, both scaling factors are fixed (Equations 7.3 and 7.4).

To identify the environmentally most beneficial process routes for C1-chemicals depending on the alternative utilization of biomass and renewable electricity, the optimization is repeated for different environmental credits for the alternative utilization of biomass (EI_{BM}) and renewable electricity (EI_{RE}).

7.3 Case study for methanol and methane production

In this section, a case study is presented for the CO_2-based and biomass-based production of methanol and methane. In the case study, only forest residue wood is considered as biomass feedstock. The objective of the considered systems is the minimization of global warming impacts. The functional unit is the utilization of 1 MWh renewable electricity and 1 MWh forest residue wood. In Section 7.3.1, the superstructure is presented. The process data is presented in the Appendix D. The wood-based processes which are newly introduced in this chapter are presented in more detail: The global warming impacts of wood-based methanol and methane production are presented in Section 7.3.2, and the global warming impact reductions of alternative wood utilization and wood-based C1-chemical production are presented in Section 7.3.3. In Section 7.3.4, the results of the case study are presented.

7.3.1 Superstructure for case study

(I) Wood-based production of methanol and methane. For the wood-based production of methanol and methane, gasification of wood is considered. Circulated fluidized bed gasifiers seem most suitable for large-scale syngas production considering throughput, cost, complexity, and efficiency issues (Hamelinck and Faaij, 2002). The gasification of wood yields syngas consisting of mainly hydrogen, carbon monoxide and CO_2. Syngas usually contains not enough hydrogen to convert all carbon monoxide and CO_2 to methane or methanol (Gassner and Maréchal, 2008). In this case, the addition of hydrogen can increase the conversion to methane or methanol. For methanol, also natural gas can be added to increase the methanol yield (Clausen et al., 2010).

Conversion of wood to methanol. For methanol production, gasification of wood with steam to syngas is considered. Subsequently, the syngas is converted to methanol at 144 bar and 235 °C. A detailed description of the process can be found in Clausen et al. (2010). For the case study, 3 wood-based processes for methanol are considered:

- a) exclusively wood is used as input,
- b) wood and hydrogen are used as inputs and
- c) wood, hydrogen and natural gas are used as inputs.

Conversion of wood to methane. For methane production, the gasification of wood is considered in an indirectly heated, steam-blown gasifier. The resulting syngas is converted to methane at 300-400 °C. A detail description of the process can be found in Gassner and Maréchal (2008, 2012). The following 2 processes were considered in the superstructure:

- a) exclusively wood is used as input and
- b) wood and hydrogen are used as inputs.

(II) CO_2-based production of methanol and methane. For both methanol and methane production, the CO_2-based processes presented in Chapter 4 are considered. For the case study, average process data for both processes is selected.

(III) Hydrogen production. For hydrogen supply, 2 options are considered: steam-methane-reforming and water electrolysis. For the hydrogen supply by electrolysis, only the utilization of renewable electricity is considered. The utilization of renewable electricity causes no environmental impacts, but competes with alternative utilization options of renewable electricity. The hydrogen supply by steam-methane-reforming is not constrained. The global warming impacts of hydrogen production by steam-methane-reforming are 10.6 kg CO_2-eq/kg.

(IV) Fossil-based production of methanol and methane. The fossil-based processes presented in Chapter 4 are considered. As for the CO_2-based processes, average process data for both fossil-based processes is selected.

(V) Alternative utilization of wood and renewable electricity. An important feature of the approach is the presentation of results as a function of GW credit for alternative utilization of both renewable inputs wood and renewable electricity. For renewable electricity, the GW credit corresponds to the maximum GW reductions presented in Chapter 6. For wood, corresponding GW reductions are shown in Section 7.3.3.

Supply processes

Renewable feedstocks: wood and renewable electricity. For the optimization performed in Section 7.3.4, the environmental impact for the supply of the renewable inputs wood and renewable electricity is not considered. For wood, only the CO_2 uptake during the growth phase is considered. This corresponds to the procedure for the determination of the maximum environmental impact reduction. For the hot spot analysis of wood-based chemicals presented in Section 7.3.2, a global warming impact for wood supply of -0.42 t CO_2-eq/MWh wood is considered (Gassner and Maréchal, 2008). This value considers chopping, transportation and CO_2 uptake during the growth phase.

CO_2 supply. For global warming impacts of CO_2 supply, a range from -0.42 to -0.99 kg CO_2-eq/kg is considered (cf. Chapter 4).

Utility supply. For the natural gas supply, the global warming impact of the EU-27 supply mix is considered. For the electricity supply, the average EU-27 grid mix in 2020 is considered. This data is applied for all processes with electricity demand except electrolysis. The electrolysis uses renewable electricity.

7.3.2 Global warming impacts for Wood-to-Methanol and Wood-to-Methane

In this section, the global warming impacts of wood-based methanol and methane production are presented. The results are presented per MWh wood used. For CO_2-based methanol and methane, the global warming impacts can be found in Section 4.3.

Methanol. The global warming impacts of wood-based and fossil-based methanol production are shown in Figure 7.2. All considered wood-based processes achieve lower global warming impacts than the fossil-based process.

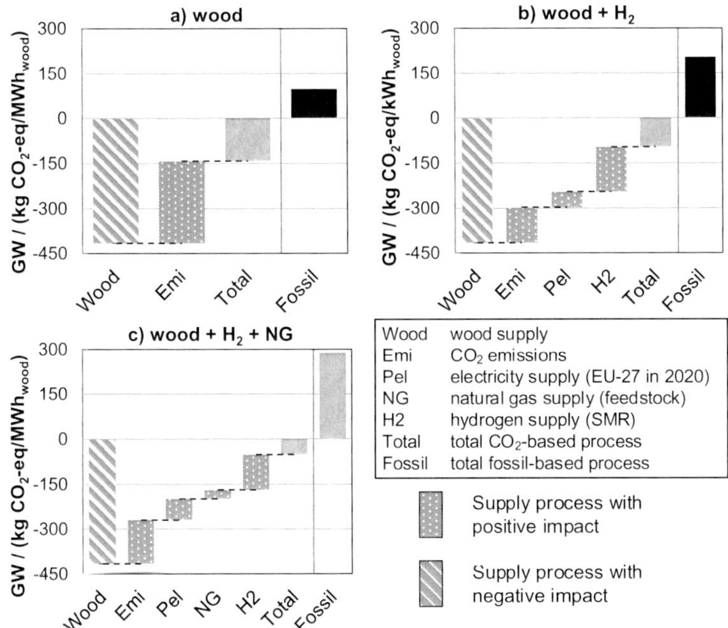

Figure 7.2: Break down of global warming impacts for methanol from wood. The reference process is fossil-based methanol production.

If wood is used as only input (Figure 7.2a), the total global warming impact is -0.14 t CO_2-eq/MWh wood. In this case, the global warming impact is exclusively due to wood supply and CO_2 emissions. About 35 % of carbon in wood is converted to methanol and about 65 % of carbon is emitted in the form of CO_2. From 1 MWh wood, 103 kg methanol can be produced. Thus, the global warming impact per kg methanol is −1.26 kg CO_2-eq.

If wood is converted together with hydrogen, the total global warming impact is −0.096 t CO_2-eq/MWh wood. This process has higher global warming impacts per MWh wood than the process that only uses wood. The higher impacts are due to the hydrogen supply. However, the process using wood and hydrogen also produces more methanol per MWh wood (216 kg methanol). The global warming impact per kg methanol is −0.39 kg CO_2-eq. In this process, about 70 % of carbon in wood is converted to methanol.

If wood is converted together with hydrogen and natural gas, the total global warming impact is −0.052 t CO_2-eq/MWh wood. Through the additional utilization of natural gas as carbon source, this process yields 306 kg methanol per MWh wood. About 75 % of carbon in wood and natural gas are converted to methanol. The global warming impact per kg methanol is −0.15 kg CO_2-eq.

Methane. The global warming impacts of wood-based and fossil-based methane production are shown in Figure 7.3. All considered wood-based processes achieve lower global warming impacts than the fossil-based process.

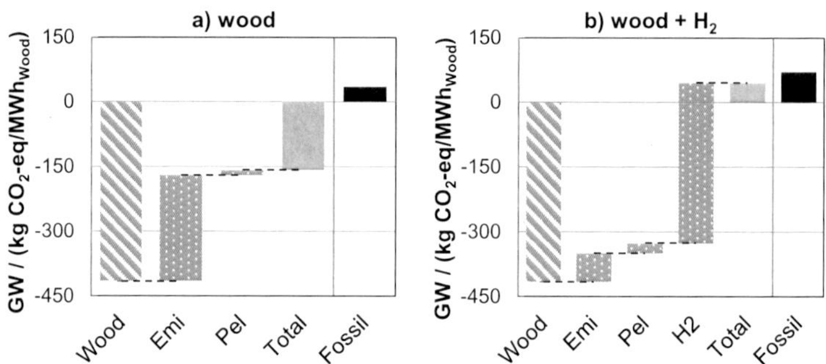

Figure 7.3: Break down of global warming impacts for methane from wood. The reference process is natural gas.

If only wood is used as input (Figure 7.3a), the total global warming impact is −0.158 t CO_2-eq/MWh wood. The majority of the impact is due to wood supply and CO_2 emissions. About 40 % of carbon in wood is converted to methane and about

60 % of carbon is emitted in the form of CO_2. This process yields 0.86 MWh methane per MWh wood. The global warming impact per MWh methane is −0.184 t CO_2-eq.

If wood is converted together with hydrogen, the total global warming impact is 0.046 t CO_2-eq/MWh wood. In this case, the additional supply of hydrogen by steam-methane-reforming leads to a positive global warming impact. In this process, about 84 % of carbon in wood is converted to methane. The methane yield is 1.77 MWh per MWh wood. The global warming impact per MWh methane is 0.026 t CO_2-eq.

7.3.3 Comparison of maximum global warming impact reductions for Wood-to-Methanol and Wood-to-Methane to alternative wood utilization

In this section, different utilization options are investigated for the renewable input wood. For this purpose, the maximum global warming impact reductions are determined per MWh wood used. This comparison is similar to the approach presented in the Chapter 6 for renewable electricity.

Process data. In the following, the processes considered are mentioned for the non-chemical utilization of wood.

Wood-to-Power. For conversion of wood to power, a range of efficiencies from 25 % to 35 % is considered (Steubing et al., 2012).

Wood-to-Heat. For conversion of wood to heat, a range of efficiencies from 68 % to 85 % is considered (Steubing et al., 2012).

Wood-to-Mobility. For the utilization of wood for mobility, ethanol production from wood and subsequent utilization of ethanol as fuel is considered. The data is taken from Steubing et al. (2012).

Conventional processes. The same conventional processes as in Chapter 6 are considered. Additionally, the substitution of base-load power plants such as coal-fired power plants is considered because biomass can be supplied continuously.

Results. Figure 7.4 compares GW impact reductions of methane and methanol production with the alternative utilization of biomass. For processes with a hydrogen demand, GW reduction is shown for hydrogen supply by steam-methane-reforming and for hydrogen supply with a global warming impact of zero ("free" hydrogen).

If the global warming impact for hydrogen supply is zero, methanol production achieves highest global warming impact reduction followed by methane production. In this case, the production of methanol and methane outperform all non-chemical

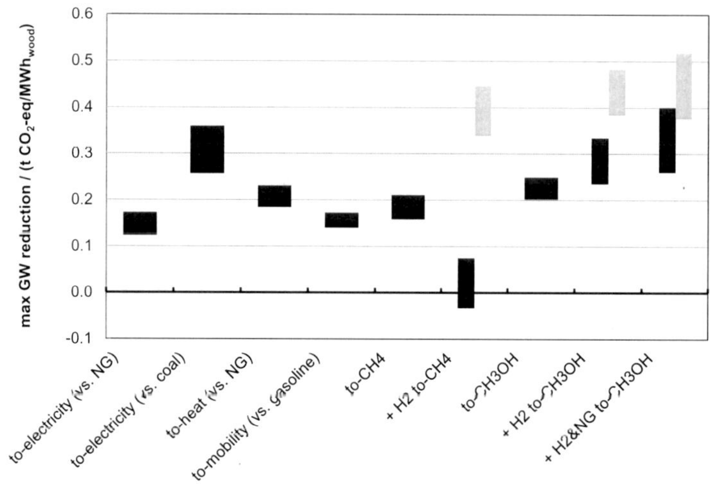

Figure 7.4: Maximum global warming (GW) impact reduction for biomass routes per MWh biomass. For the gray bars, the global warming impact of hydrogen supply is not considered.

utilization options (power, heat and mobility). Among the non-chemical utilization options, Wood-to-Power (vs coal) achieves highest global warming impact reductions (0.30–0.35 t CO_2-eq/MWh). The global warming impact reductions of the other non-chemical utilization options are in the range of 0.15–0.25 t CO_2-eq/MWh.

If hydrogen is supplied by steam-methane-reforming, global warming impact reductions for methane and methanol decrease. In this case, Wood-to-Power has highest impact reduction if coal electricity is substituted. However, Wood-to-Methanol and Wood-to-Methane are both still competitive to other non-chemical utilization options.

For methanol production, the global warming impact reduction per MWh wood is always higher for the combined utilization of wood and hydrogen than the exclusive utilization of wood.

The Wood-to-Chemical routes (cf. Figure 7.4) are more competitive to Wood-to-Alternative (i.e., power, heat and mobility) than Power(+CO_2)-to-Chemical routes (cf. Figure 6.4) to Power-to-Alternative (i.e., power, heat and mobility). This is due to the lower conversion efficiency of Wood-to-Alternative than for Power-to-Alternative. The routes Wood-to-Power and Wood-to-Mobility are constrained by the Carnot-efficiency in contrast to Power-to-Power and Power-to-Mobility. Also the Wood-to-Heat route has a lower efficiency than the Power-to-Heat route if a heat pump is used for Power-to-Heat.

7.3.4 Comparison of CO_2 and biomass as feedstock for C1-chemicals

In this section, the results of the optimization are presented for the production of methanol and methane from renewable inputs. Based on the optimization, it is identified under which conditions the renewable inputs wood and renewable electricity should be used for production of methanol or wood (Figure 7.5). Here, the conditions are expressed as the global warming credits (GW credits) for the alternative utilization of wood and renewable electricity.

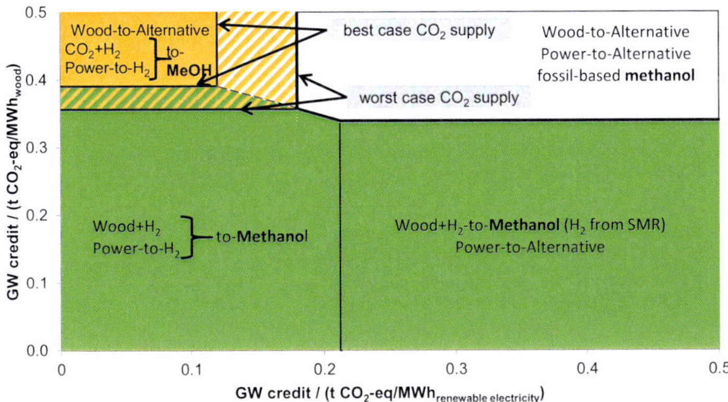

Figure 7.5: Optimal process routes for methanol and methane production depending on the GW credit for Power-to-Alternative and Wood-to-Alternative. Methane is always produced conventionally. For the striped-areas, the optimal process routes additionally depends on the global warming impact of CO_2 supply.

In the case study, the production of methanol from renewable carbon sources always achieves lower global warming impacts than the production of methane. Thus, methane is never produced from biomass and renewable electricity in this case study. Methanol should be produced from wood if the GW credit for alternative utilization of wood is lower than 0.35 t CO_2-eq/MWh. Among the wood-based processes for methanol, the combined utilization of wood and hydrogen achieves the lowest global warming impacts (see Figure 7.4). If the GW credit for alternative utilization of renewable electricity is higher than 0.22 t CO_2-eq/MWh, the hydrogen should be produced by steam-methane-reforming; the renewable electricity should be used alternatively. If the GW credit for alternative utilization of renewable electricity is lower than 0.22 t CO_2-eq/MWh, the renewable electricity should be used for hydrogen production.

Assuming the best case for CO_2 supply (orange & orange-striped area in Figure 7.5, GW CO_2 supply = −0.99 kg CO_2-eq/kg), methanol should be produced from CO_2 if the GW credit for alternative utilization of renewable electricity is lower than 0.17 t CO_2-eq/MWh and the GW credit for alternative utilization of wood is higher than 0.36 t CO_2-eq/MWh (area with orange box and orange-stripped box). For the worst case of CO_2 supply (orange area, GW CO_2 supply = −0.42 kg CO_2-eq/kg), the GW credit for alternative utilization of renewable electricity must be lower than 0.12 t CO_2-eq/MWh and the GW credit for alternative utilization of wood must be higher than 0.39 t CO_2-eq/MWh, before CO_2 should be used to produce methanol.

7.4 Conclusions

In this chapter, an approach based on optimization is presented for comparison of biomass-based and CO_2-based processes for C1-chemicals. The particularity of CO_2-based and biomass-based processes is that both rely on renewable inputs: biomass and renewable electricity. Thus, the comparison takes into account the potential alternative utilization of biomass and renewable electricity.

Due to the different types of biomass with different characteristics such a comparison for CO_2-based processes must be conducted for each biomass type. If biomass (instead of biomass residues) is compared to CO_2-based processes, the functional unit for biomass could also be the cultivation of a specific area (e.g., 1 m^3). In this case, also different types of biomass could be compared to CO_2-based processes.

In this chapter, the approach is presented for a case study that compares the utilization of forest residue wood for methanol and methane with CO_2-based methanol and methane production. The results of the case study show that wood-based methanol production is very promising if additionally hydrogen is used as input. Among the considered alternative utilization options for wood, only the conversion of wood to electricity achieves higher global warming impact reductions if a coal-fired power plant is replaced.

For methanol production, the CO_2-based processes are only beneficial compared to wood-based processes if the GW credit for Wood-to-Alternative is higher than 0.36 t CO_2-eq/MWh. However, among the Wood-to-Alternative options considered, only the best case conversion of wood to electricity with substitution of coal-fired power plants can achieve such a high GW credit. Thus, for the production of methanol, wood is the preferred renewable carbon source in the case study presented.

However, due to missing data, the alternative utilization of wood for other chemicals (with more than 1 carbon atom) is not considered in this chapter. These processes have the potential to be even more beneficial than methanol, in particular, if fossil-based processes with several productions steps and thus high environmental impacts are replaced.

Nonetheless, the main insight gained in this chapter is that the methanol production from wood together with renewable electricity achieves higher benefits for global warming impacts than the utilization of CO_2 together with renewable electricity. Thus, future research should also assess the utilization of biomass together with renewable electricity for the production of chemicals with more than 1 carbon atom.

Chapter 8

Summary and conclusions

The conversion of CO_2 to C1-chemicals such as formic acid, carbon monoxide, methane and methanol is a currently often discussed route for the integration of renewable electricity into the chemical industry. Formic acid, methanol and methane can additionally be used as portable and dispatchable energy carriers. For methanol and methane, existing LCA studies already show that CO_2-based production of C1-chemicals is environmentally beneficial compared to fossil-based processes if hydrogen is supplied by renewable electricity. Currently, renewable electricity is limited. Thus, the assessment of CO_2-based C1-chemicals should consider alternative utilization options for renewable electricity. However, such a system-wide comparison has been missing till now. In this thesis, an approach is introduced that allows to compare CO_2-based C1-chemicals, and additionally benchmarks CO_2-based C1-chemicals with alternative utilization options for renewable electricity.

Approach — Maximum environmental impact reduction

In this thesis, CO_2-based C1-chemicals are compared based on the maximum environmental impact reduction. The environmental impact reduction is the difference of the environmental impact of CO_2-based and corresponding fossil-based processes. Thus, the ideal CO_2-based process would have very low environmental impacts and would replace a fossil-based process with very high environmental impacts. Previous work has shown that the environmental impact of hydrogen supply is crucial for CO_2-based C1-chemicals. Thus, the maximum environmental impact reduction is obtained if the environmental impact of hydrogen supply is not considered. The maximum environmental impact reduction is presented per kg hydrogen or per MWh electricity used for electrolysis depending on the scope of the chapter. By presenting the results per kg hydrogen, the maximum environmental impact reduction is identical with the break-even environmental impact of hydrogen supply. In other words, if hydrogen is supplied with lower environmental impacts than the value for the maximum environmental impact reduction, the CO_2-based process reduces environmental impacts compared to the fossil-based process. Thus, this approach allows to identify suitable hydrogen supply processes for environmentally beneficial CO_2-based processes. Additionally, the maximum environmental impact reduction also reveals which CO_2-based process achieves the highest environmental impact reductions for a particular hydrogen supply process, e.g., hydrogen supply by electrolysis using wind electricity. This information is in particular relevant if the considered hydrogen supply is constrained. By presenting the maximum environmental impact reduction per MWh renewable electricity, all processes using renewable electricity can be compared and ranked. Hence, decision makers can identify processes that use renewable electricity most efficiently in environmental terms.

CO_2-based C1-chemicals

Utilization of CO_2-based C1-chemicals in the chemical industry. The considered C1-chemicals are important feedstocks (methane) and intermediates (formic acid, carbon monoxide and methanol). Among the considered C1-chemicals from CO_2, formic acid achieves the highest maximum environmental impact reduction. The production of formic acid can reduce environmental impacts compared to fossil-based processes even if hydrogen is supplied by fossil-based steam methane reforming. The CO_2-based production of carbon monoxide and methanol can also be environmentally beneficial if hydrogen is supplied by fossil-based steam methane reforming. However, both carbon monoxide and methanol require an efficient supply of CO_2 and the substitution of a low-efficiency fossil-based process. Otherwise, the supply of hydrogen by renewable electricity is required to achieve environmental impact reductions compared to fossil-based processes. For the CO_2-based production of methane, hydrogen supply by renewable electricity is mandatory to reduce environmental impacts compared to fossil-based processes.

The high environmental impact reduction of CO_2-based formic acid is primarily due to the substitution of an energy-intensive fossil-based process. The fossil-based process consists of 5 process steps: (i) production of syngas from natural gas, (ii) separation of carbon monoxide, (iii) conversion of carbon monoxide with methanol to methyl formate, (iv) conversion of methyl formate and water to formic acid and methanol, and finally (v) separation of formic acid. In contrast, the CO_2-based production of methane competes with natural gas that only needs to be purified after extraction from underground deposits.

CO_2-based C1-chemicals as energy carriers. Among the considered CO_2-based C1-chemicals, formic acid, methanol and methane are frequently discussed as energy carriers. However, the environmentally beneficial supply of CO_2-based energy carriers is more challenging than the utilization of CO_2-based chemicals in the chemical industry: While the environmental impact of the CO_2-based process is independent from the utilization as energy carrier or chemicals, the environmental impact of the substituted fossil-based processes depends on the utilization as energy carriers or as chemical. Usually, the environmental impact of fossil-based energy carriers is lower than chemicals because current energy carriers are almost directly taken from underground deposits. If CO_2-based C1-chemicals are used as energy carriers, the substituted fossil-based process (e.g., natural gas) is equal for all CO_2-based C1-chemicals. In this case, the environmental impact of the CO_2-based C1-chemical strongly affects the order of the environmental impact reductions for CO_2-based energy carriers. Consequently, the most promising CO_2-based energy carrier is methane, because the production of CO_2-based methane has lowest environmental impacts per MWh lower heating value among the CO_2-based process.

Insights from the system-wide perspective.

Hydrogen supply. In this thesis, most CO_2-based processes for C1-chemicals require hydrogen as co-reactant. Only the dry reforming of methane (DRM) reaction does not require hydrogen as co-reactant. However, the DRM reaction requires additional hydrogen to substitute the fossil-based steam-methane-reforming, which co-produces more hydrogen than the DRM reaction. For all considered CO_2-based processes except formic acid, the hydrogen must be supplied by renewable electricity to achieve lower environmental impacts than the fossil-based processes, independently from the efficiency of the fossil-based process. Currently, the hydrogen production from renewable electricity competes with alternative utilization options for renewable electricity. To use renewable electricity most efficiently in environmental terms, renewable electricity should be employed in technologies in the order of highest environmental impact reductions. According to Chapter 6, highest global warming and fossil depletion impact reductions are achieved by using renewable power in heat pumps (Power-to-Heat) and battery electric vehicles (Power-to-Mobility). The third highest environmental impact reductions are achieved through the substitution of fossil-based power plants (Power-to-Power). Significant lower environmental impact reductions are achieved for the production of hydrogen (Power-to-H_2) and the subsequent substitution of fossil-based hydrogen production. Among the considered utilization options for renewable electricity, the lowest environmental impact reductions are achieved for CO_2-based production of C1-chemicals.

In summary, CO_2-based processes for C1-chemicals can be used to integrate renewable electricity into the chemical industry and reduce environmental impacts. However, from an environmental point of view, renewable electricity should not be used for production of C1-chemicals until utilization options for renewable electricity are exhausted in other sectors. Currently, CO_2-based processes for C1-chemicals are most promising in remote areas with low residential and industrial energy demands, and where an export of renewable electricity via electricity interconnectors is not possible.

CO_2 supply. For the CO_2 supply, various sources are considered including capture from flue gases and air. The highest environmental impact reductions are achieved for each CO_2-based process if CO_2 is captured from flue gases. If CO_2 is captured from the air, environmental impact reductions decrease. The lower CO_2 concentration in the air compared to flue gases results in an increased energy demand for CO_2 capture. However, air capture is more environmentally beneficial than CO_2 capture from flue gases if the operation of the process that supplies the flue gas is exclusively a consequence of the CO_2 utilization, e.g., additionally installed fossil-based power plant with CO_2 capture that supplies CO_2 and electricity to the CO_2-based process. A further example is the avoided installation of renewable electricity as a consequence of CO_2 capture from fossil-based power plants. If captured CO_2 from flue gases is used for CO_2-based processes that otherwise would be stored in the underground (CCS), air capture achieves higher global warming impact reductions but lower fossil depletion impact reductions compared to CO_2 capture from flue gases.

Alternative renewable production of C1-chemicals. Besides CO_2, biomass is a further renewable carbon source for the chemical industry. In contrast to CO_2, biomass is limited. Currently, biomass is primarily considered for generation of electricity and as automotive fuel. Since biomass is limited, also a comparative assessment for the biomass utilization options is required. Such a comparison for biomass was presented by Steubing et al. (2012). However, the utilization of biomass as chemical feedstock is missing in the analysis. In Chapter 6, an approach for the comparison of biomass and CO_2 as feedstock for the chemical industry is presented. For the presented case study, biomass seems more promising as feedstock than CO_2 if the alternative utilization of limited renewable electricity and biomass is considered. Furthermore, the utilization of hydrogen from limited renewable electricity is more promising for upgrading biomass routes than for CO_2-based processes.

Flexibilization of electricity supply and demand. In the past, the fixed electricity demand was supplied by dispatchable fossil-based resources. Through the increasing penetration of intermittent renewable electricity, a mix of dispatchable demand and supply is required to match demand and supply. The production of CO_2-based chemicals can be used to absorb surpluses in electricity supply if hydrogen is produced in a flexible electrolysis. The hydrogen can be stored, e.g., in caverns and subsequently converted with CO_2 to C1-chemicals.

If CO_2-based C1-chemicals are used as energy carriers, they can also be reconverted to electricity in times of electricity shortage. CO_2-based C1-chemicals are suited for long-term storage, because of the high energy density (methanol and formic acid) and the already existing storage capacities (methane).

In an energy system based on 100 % renewable energy, CO_2-based C1-chemicals will be probably required as energy carriers. However, since the production of environmentally beneficial CO_2-based energy carriers is very challenging the amount required should be as low as possible. An alternative dispatchable source for electricity production, which is also based on CO_2-based C1-chemicals is for example the combustion of materials derived from CO_2-based C1-chemicals in waste incineration power plants at their end-of-life.

Future perspective for CO_2-based chemicals. The production of C1-chemicals from CO_2 is one option to integrate renewable electricity into the chemical sector and subsequently substitute fossil-based processes. The considered C1-chemicals are predominantly used at the start of current value chains for chemical products. Thus, the potential to reduce fossil-based processes is not fully exploited for the new feedstock CO_2. The results for the CO_2-based C1-chemicals already show that formic acid is most promising because the fossil-based process has the highest environmental impacts among all considered fossil-based processes. A further promising process is the CO_2-based polyol production (von der Assen and Bardow, 2014). Through the utilization of CO_2, the demand of fossil-based epoxides can be reduced. The CO_2-based polyol production does not require hydrogen and thus no renewable electricity.

For the further identification of promising CO_2-based processes, a systematic screening is required for CO_2 utilization options. This screening should in particular search for CO_2-based processes where the number of reaction steps can be decreased compared to fossil-based processes. Ideally, a CO_2-based chemical can be produced in 1-step and replaces a product at the end of the current value chain for chemicals. Per carbon atom, products at the end of a value chain always have higher environmental impacts than intermediates and feedstocks because the environmental impacts accumulate during the production processes.

An alternative option to substitute chemicals or materials that can not be produced from CO_2 is the production of tailor-made materials from CO_2 with similar or even better properties than currently used fossil-based materials. In LCA language, a CO_2-based service substitutes a fossil-based service.

Appendix A

Process data for C1-chemicals

A.1 Input and output data for CO_2-based processes

In the following flowsheets (A.1-A.18), the amount of reactants, products and emissions is shown for all processes considered. The lower heating value (LHV) of inputs and outputs is shown in parentheses. Additionally, the electricity and heat demand is presented. For all processes, the electricity demand relates to hydrogen and CO_2 inlet pressures presented in the corresponding references. The additional electricity demand that is required if hydrogen and CO_2 are supplied at 1 bar is presented in parentheses. Please note that the energy balance is not closed because for most processes the cooling demand was not reported.

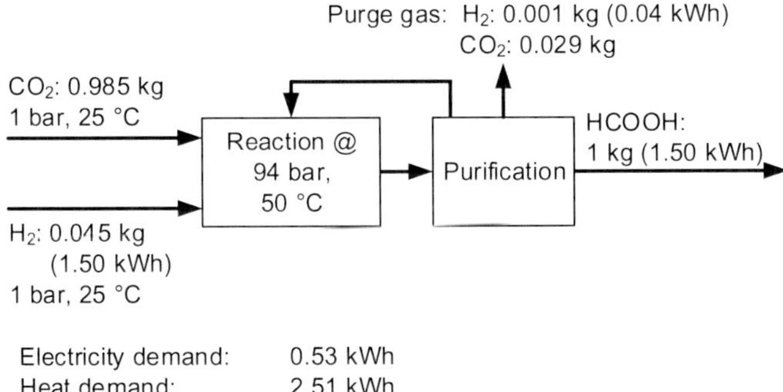

Figure A.1: Flowsheet for CO_2-based formic acid production with process data from Jens et al. (2017)

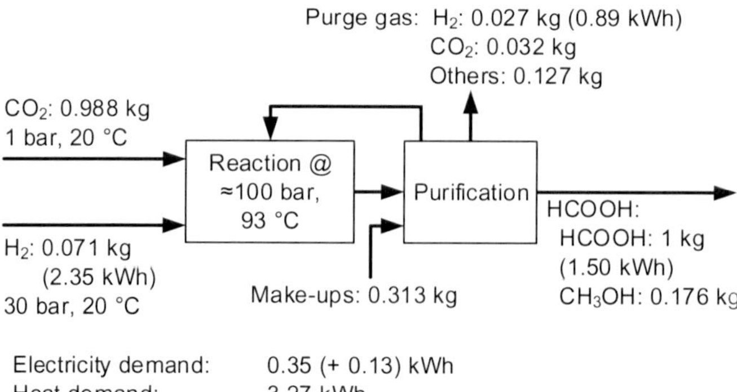

Figure A.2: Flowsheet for CO_2-based formic acid production with process data from Pérez-Fortes et al. (2016)

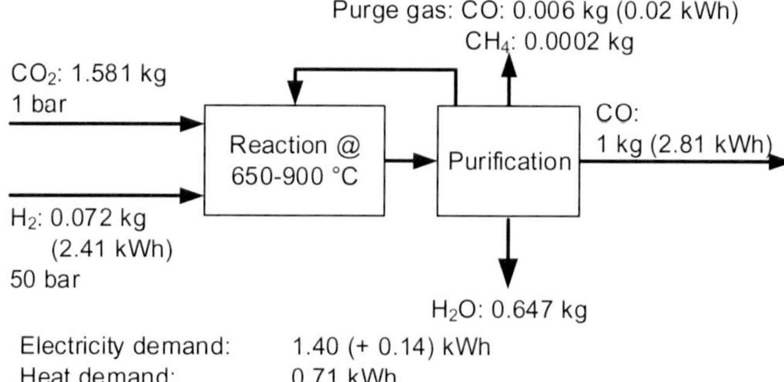

Figure A.3: Flowsheet for CO_2-based carbon monoxide production via rWGS with process data from CO_2RRECT (2014)

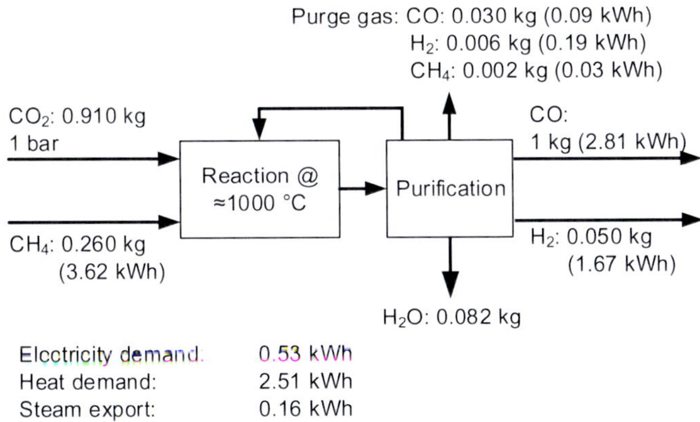

Figure A.4: Flowsheet for CO_2-based carbon monoxide production via DRM with process data from CO_2RRECT (2014)

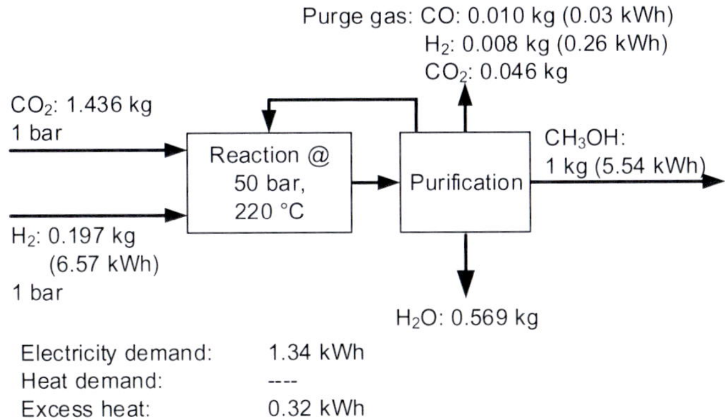

Figure A.5: Flowsheet for CO_2-based methanol production with process data from Rihko-Struckmann et al. (2010)

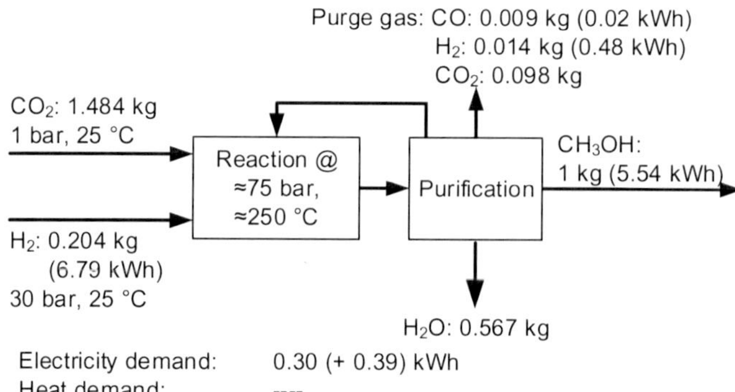

Figure A.6: Flowsheet for CO_2-based methanol production with process data from Van-Dal and Bouallou (2013)

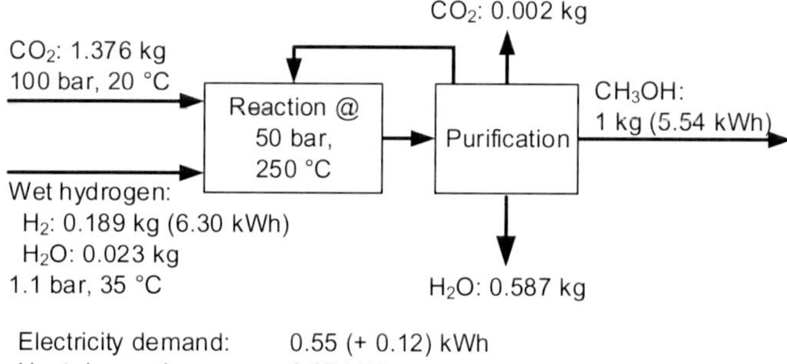

Figure A.7: Flowsheet for CO_2-based methanol production with process data from Kiss et al. (2016)

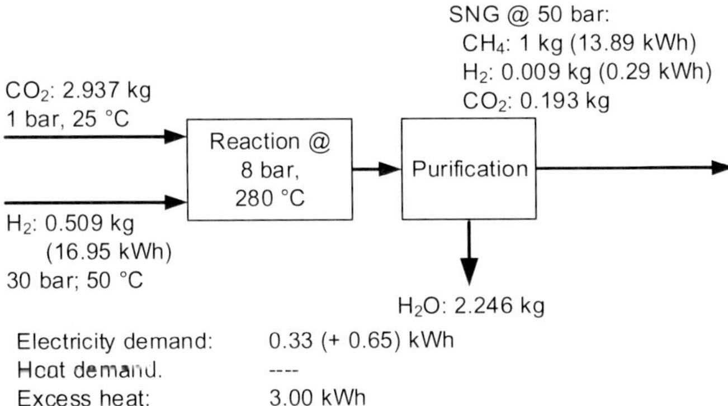

Figure A.8: Flowsheet for CO_2-based methane production with process data from Müller et al. (2011)

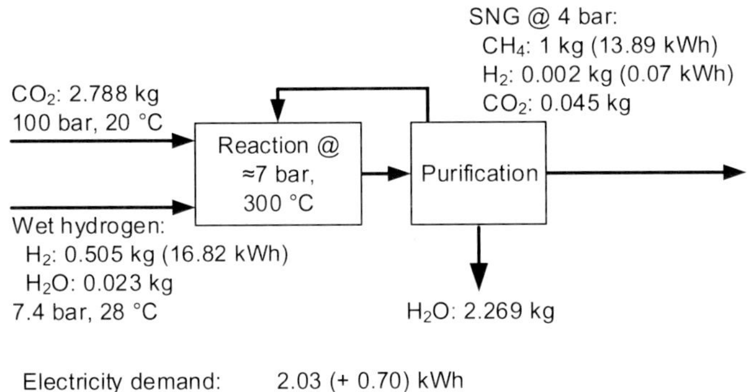

Figure A.9: Flowsheet for CO_2-based methane production with process data from Jean et al. (2014)

A.2 Input and output data for fossil-based processes

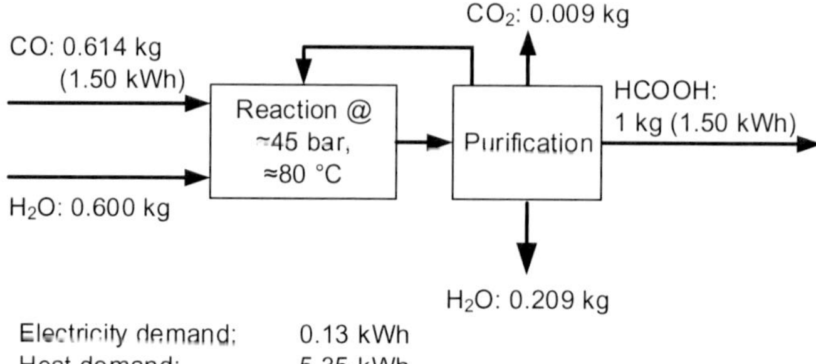

Figure A.10: Flowsheet for fossil-based formic acid production with process data from ecoinvent (2007c). The purity of formic acid is not mentioned in the reference. The carbon balance indicates that results are presented per kg pure formic acid.

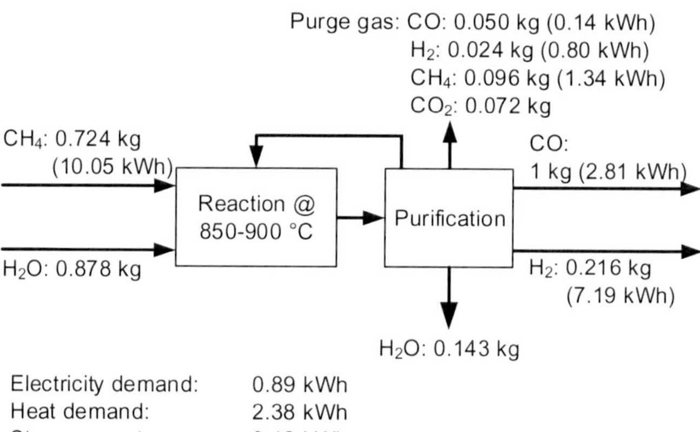

Figure A.11: Flowsheet for fossil-based carbon monoxide production with process data from CO_2RRECT (2014). The molar H_2/CO ratio is 3.

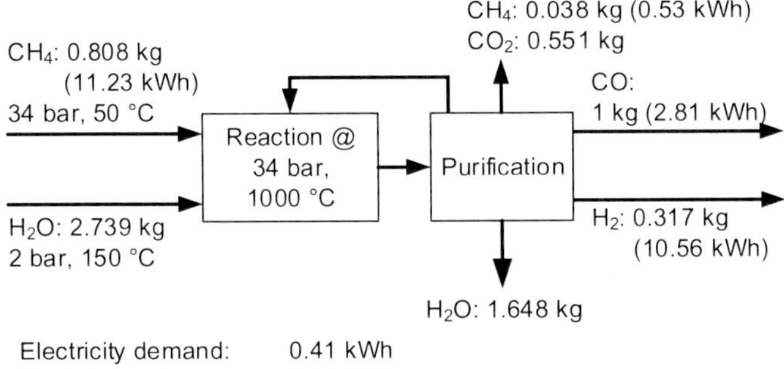

Figure A.12: Flowsheet for fossil-based carbon monoxide production with process data from Baltrusaitis and Luyben (2015). The molar H_2/CO ratio is 4.4.

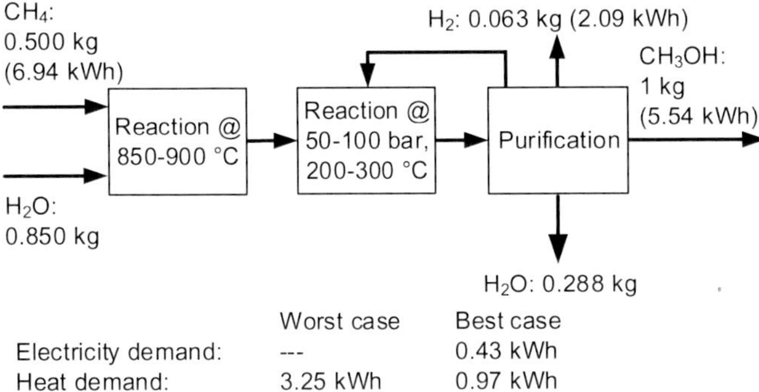

Figure A.13: Flowsheet for fossil-based methanol production with process data from ecoinvent (2007a). The electricity and heat demand consider best and worst case values.

A.3 Input and output data for biomass-based processes

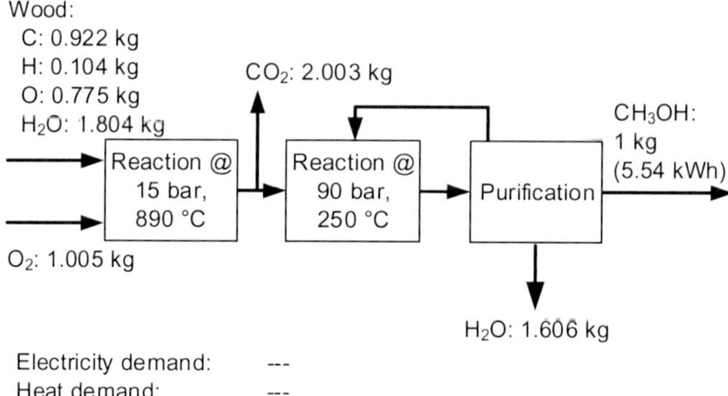

Figure A.14: Flowsheet for wood-based methanol production with process data from Hamelinck and Faaij (2002). The wood is dried before it is fed into the reactor.

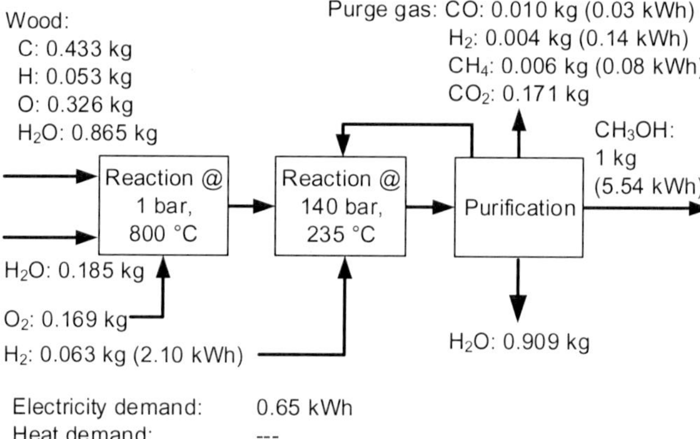

Figure A.15: Flowsheet for wood-based methanol production with process data from Clausen et al. (2010). The wood is dried before it is fed into the reactor.

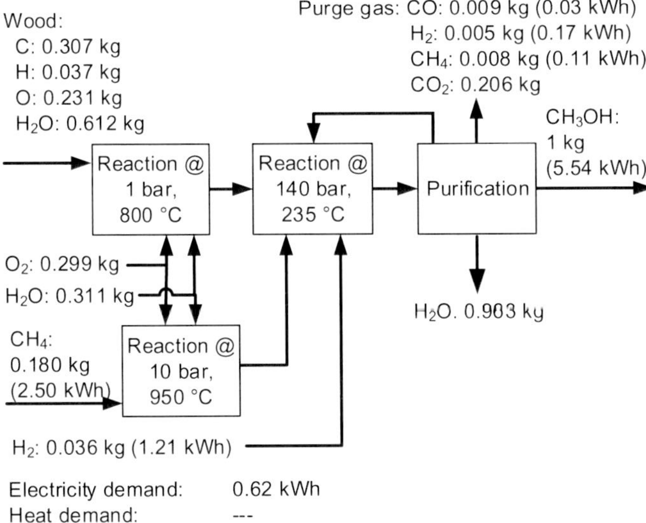

Figure A.16: Flowsheet for wood-based methanol production with process data from Clausen et al. (2010). The wood is dried before it is fed into the reactor.

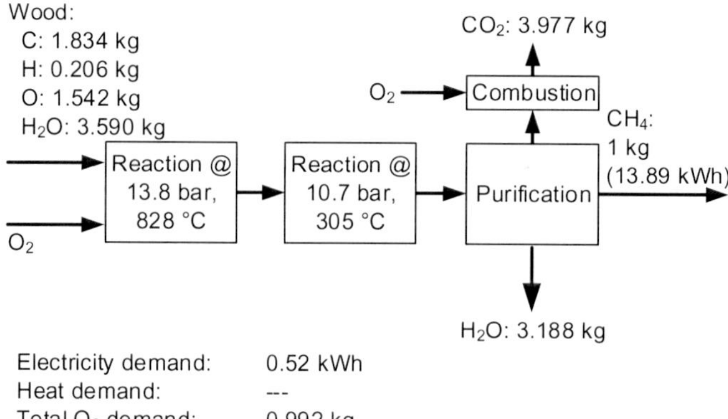

Figure A.17: Flowsheet for wood-based methane production with process data from Gassner and Maréchal (2008). The wood is dried before it is fed into the reactor.

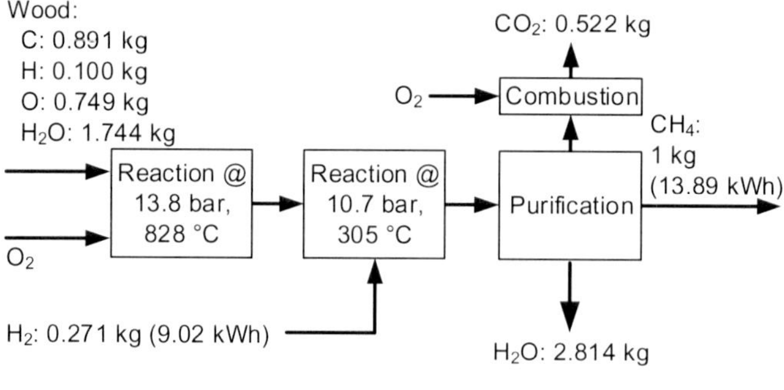

Figure A.18: Flowsheet for wood-based methane production with process data from Gassner and Maréchal (2008). The wood is dried before it is fed into the reactor.

A.4 CO_2 capture

Table A.1: Energy demands for supply of the product CO_2 at 100 bar; data summarized from von der Assen et al. (2016)

CO_2 source	Energy demands in GJ per t CO_2			
	Electricity	Heat	NG	Coal
Wet air capture	0.37–1.58	–	4.19–7.67	–
NGCC power plant*	1.22–2.38	–	–	–
IGCC power plant	0.25–1.07	–	0–1.22	–
Coal power plant*	0.76–1.84	–	–	–
Refineries and steam cracker				
with post combustion	0.48–0.55	3.85–4.21	–	–
with pre combustion	−0.10	–	4.50	–
with oxyfuel	−8.40–2.00	–	0–12.50	–
Integr. pulp and paper mill	0.04–1.09	–	0–1.57	–
with additional CHP	−2.48	–	4.50	–
Market pulp mill	0.24–1.29	–	0–1.00	–
Iron and steel plant				
w/ air-blown blast furnace	0.54–0.91	0–4.40	–	–
w/ O_2-blown blast furnace	0.62–1.50	0–3.30	–	–
with COREX smelter	0.45–0.97	0–4.40	–	–
Cement plant				
with post combustion	0.54–0.73	2.70–3.70	–	–
with additional CHP	−0.39–(−0.35)	–	–	4.00–6.12
with oxyfuel kiln	−0.44–0.99	–	–	0.06–4.49
with calcium looping	−0.49–0.54	–	–	1.60–4.82
Ammonia plant	0.40	0.01	–	–
Ethylene oxide plant	0.40	0.01	–	–
Gas processing	0.40	0.01	–	–
Hydrogen plant	0.25–0.52	0–0.01	–	–

*with post combustion, w/ – with, NG – natural gas

Appendix B

LCA data

Table B.1: Considered LCA data sets for utility supply; database: GaBi ts (2016)

Product	Name of data set	Year	Chapter
Electricity	electricity mix 2020 (average power plant) [EU-27]	2015	4–7
Electricity	electricity mix 2030 (average power plant) [EU-27]	2015	5
Heat	steam from natural gas ($\eta = 90$ %) [EU-27]	2012	4–7
Heat	steam from natural gas ($\eta = 90$ %) [NL]	2012	4
Heat	steam from natural gas ($\eta = 90$ %) [PT]	2012	4
Natural gas	natural gas mix [EU-27]	2012	4–7
Natural gas	natural gas mix [NL]	2012	4
Natural gas	natural gas mix [PT]	2012	4

Table B.2: Considered LCA data sets for CO_2 supply from coal-fired power plant; databases: GaBi – GaBi ts (2016), eco – ecoinvent Data V 2.2 (2010)

Product	Name of data set	Year	Database	Chapter
Heat	thermal energy from hard coal [EU-27]	2012	GaBi	5 & 6
MEA	market for monoethanolamine [GLO]	2015	eco	5

Table B.3: Considered LCA data sets for construction; database: ecoinvent Data V 2.2 (2010)

Product	Name of data set	Year	Chapter
Electrolysis	PEM fuel cell 2kWe, future*	2015	5
Steel	steel production, electric, chromium steel 18/8 [RER]	2015	5

*original data set refers to fuel cell for combined heat and power (CHP); for inventory of electrolysis heat distribution system is not considered

Table B.4: Considered LCA data sets for fossil-based processes; database: GaBi ts (2016)

Product	Name of data set	Year	Chapter
Power from gas turbine	based on heat from natural gas (η = 40 %)		6 & 7
Power from coal	thermal energy from hard coal [EU-27]		7
Space heating	thermal energy from natural gas [EU-27] (η = 91 %)	2012	6 & 7
Hydrogen [b]	hydrogen (steam reforming - natural gas) [DE]	2012	6 & 7
Natural gas	natural gas mix [EU-27]	2012	6 & 7
Methanol [b]	methanol from natural gas (integrated technologies)	2012	6
Syngas [b]	syngas (H_2:CO = 3:1) from natural gas	2012	6
diesel mix	at refinery [EU-27]	2012	6
gasoline mix	RON 95, at refinery [EU-27]	2012	6 & 7

[b] country specific values derived from natural gas data set;

Table B.5: Considered LCA data sets for electricity supply scenarios (only considered for electrolysis); database: GaBi ts (2016)

Product	Name of data set	Year	Chapter
Electricity	electricity from photovoltaics [DE]	2012	4 & 5
Electricity	electricity from wind power [EU-27]	2012	4 & 5
Electricity	electricity mix 2020 (average power plant) [EU-27]	2015	4 & 5
Electricity	electricity mix 2040 (average power plant) [EU-27]	2015	4 & 5
Electricity	electricity mix 2050 (average power plant) [EU-27]	2015	4 & 5

In Chapter 6, additional country specific LCA data sets are considered. The considerd countries are presented in the Tables B.6–B.9

Table B.6: Considered countries for heat from natural gas and natural gas mix (only Chapter 6)

Continent	Number of considered countries	ISO-Code of considered countries
America	2	BR US
Asia	3	CN IN JP
Australia	2	AU NZ
Europe	23	AT BE CH DE DK ES FR GB GR HU IE IT LT LU LV NL NO PL PT RO SE SI SK

Table B.7: Considered countries for electricity mix (only Chapter 6)

Continent	Number of considered countries	ISO-Code of considered countries
America	2	BR US
Asia	3	CN IN JP
Australia	2	AU NZ
Europe	30	AT BE BG CH CY CZ DE DK EE ES FI FR GB GR HU IE IS IT LT LU LV MT NL NO PL PT RO SE SI SK

Table B.8: Considered countries for coal mix (only Chapter 6)

Continent	Number of considered countries	ISO-Code of considered countries
America	2	BR US
Asia	3	CN IN JP
Australia	2	AU NZ
Europe	10	AT BE DE DK FR GB IT NL NO SE

Table B.9: Considered countries for gasoline and diesel (only Chapter 6)

Continent	Number of considered countries	ISO-Code of considered countries
America	2	BR US
Asia	3	CN IN[a] JP
Australia	1	AU
Europe	2	DE GB

[a] For IN only diesel considered

Appendix C

CO_2-to-Chemical

C.1 Breakdown of fossil depletion impacts for CO_2-based processes

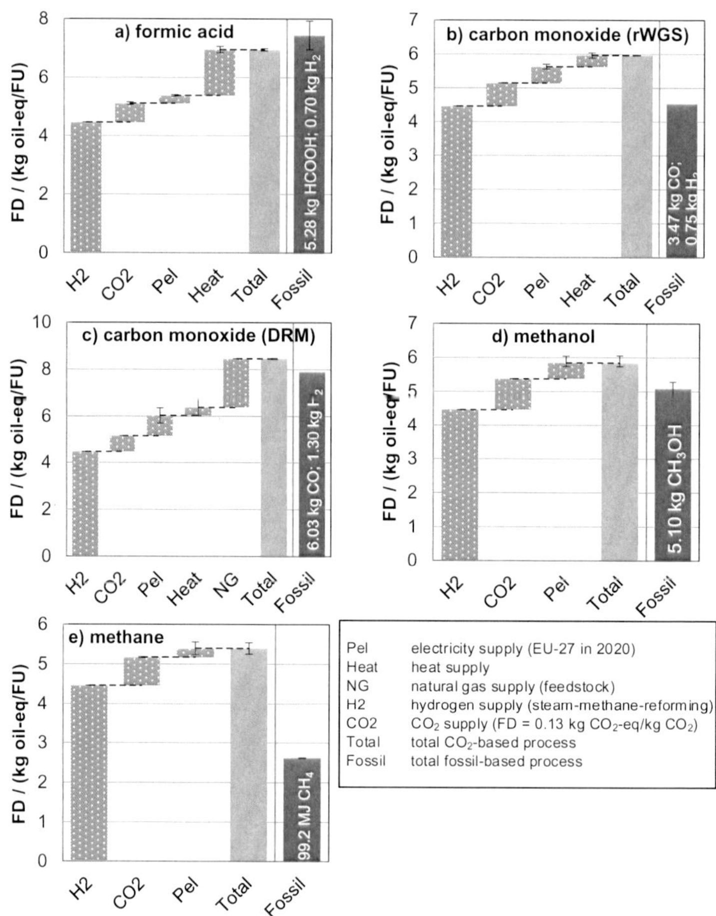

Figure C.1: Breakdown of fossil depletion (FD) impacts per functional unit (FU = use of 1 kg hydrogen) for CO_2-based processes. The error bars of the CO_2-based processes represent the range of considered process concepts. For fossil-based processes an upper bound of fossil depletion impacts (worst case) is shown. The range of the fossil-based processes is due to different yields of C1-chemicals from the considered CO_2-based processes. The average yield is shown in the right bar.

C.2 Breakdown of global warming and fossil depletion impacts for fossil processes

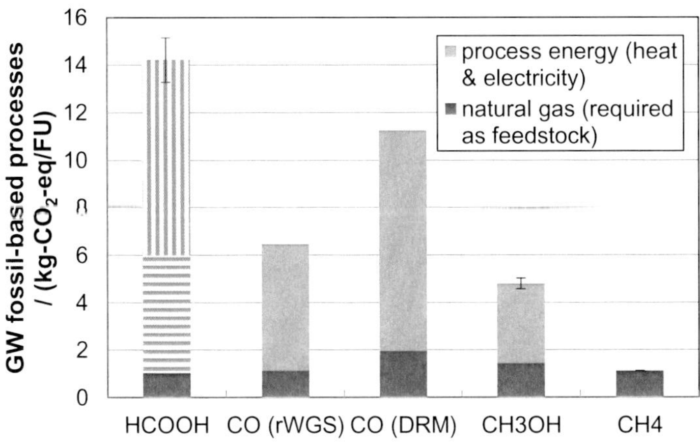

Figure C.2: Breakdown of global warming impacts (GW) for fossil-based processes per functional unit (FU). For formic acid (HCOOH), the process energy (heat and electricity) demand is divided in production of the precursor carbon monoxide (horizontal lines) and the conversion of carbon monoxide to formic acid (vertical lines). The global warming impact of NG (feedstock) includes upstream processes such as transportation and natural gas purification.

Table C.1: Yield of chemicals for considered CO_2-based processes per functional unit (use of 1 kg hydrogen)

	C1-chemical in kg	Hydrogen in kg	C1-chemical in mol	Hydrogen in mol
Formic acid	4.93–5.67	0.65–0.75	107–123	322–372
Carbon monoxide (rWGS)	3.47	0.75	124	372
Carbon monoxide (DRM)	6.03	1.30	215	645
Methanol	4.90–5.29	–	153-165	–
Methane [a]	1.98–1.99	–	123–124	–

[a] Corresponds to 98.8–99.6 MJ based on lower heating value (LHV)

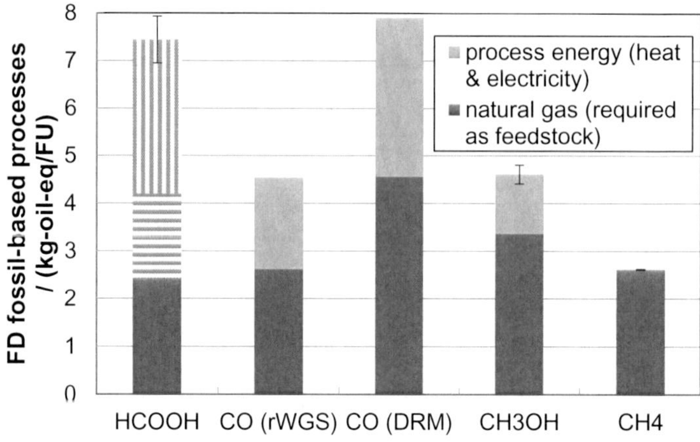

Figure C.3: Breakdown of fossil depletion impacts (FD) for fossil-based processes per functional unit (FU). For formic acid (HCOOH), the process energy (heat and electricity) demand is divided in production of the precursor carbon monoxide (horizontal lines) and the conversion of carbon monoxide to formic acid (vertical lines). The fossil depletion impact of NG (feedstock) includes upstream processes such as transportation and natural gas purification.

C.3 Maximum environmental impact reductions with alternative natural gas supply

In the main text, the maximum environmental impact reductions for global warming and fossil depletion are determined by considering natural gas supply mix in the EU-27. In this section, the maximum environmental impact reductions for global warming and fossil depletion are presented for natural gas mixes of countries with highest (Portugal) and lowest (The Netherlands) environmental impacts in the EU-27.

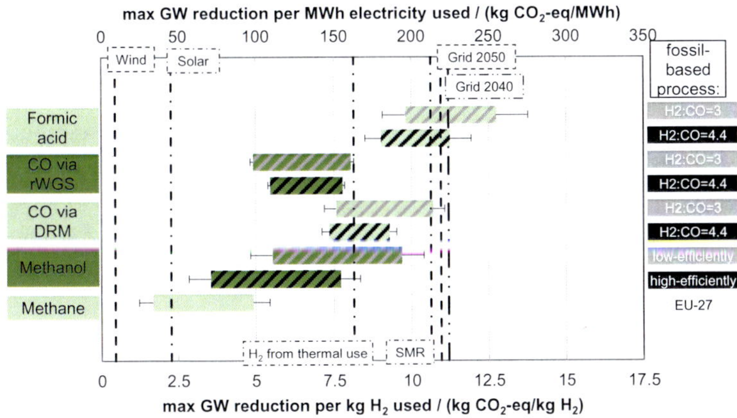

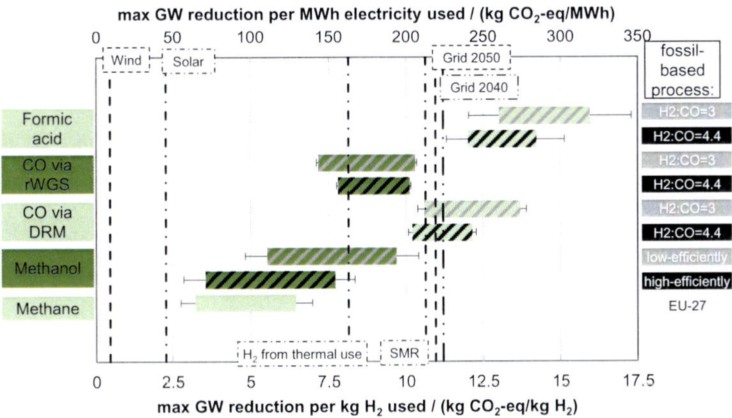

Figure C.4: Maximum global warming (GW) impact reductions for CO_2-based C1-chemicals using the Dutch (top) and Portuguese (bottom) natural gas mix. For further explanation of the Figure see caption of Figure 4.4 in main text.

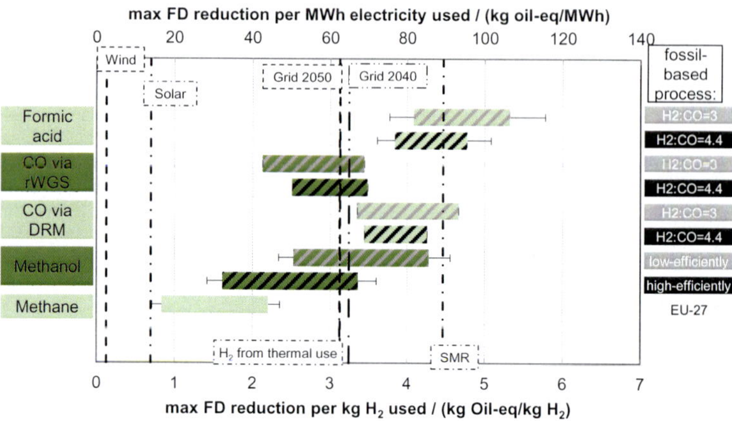

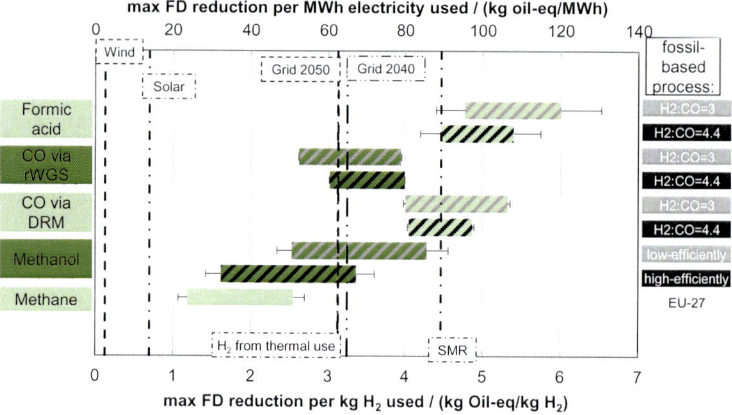

Figure C.5: Maximum fossil depletion (FD) impact reductions for CO_2-based C1-chemicals using the Dutch (top) and Portuguese (bottom) natural gas mix. For further explanation of the Figure see caption of Figure 4.4 in main text.

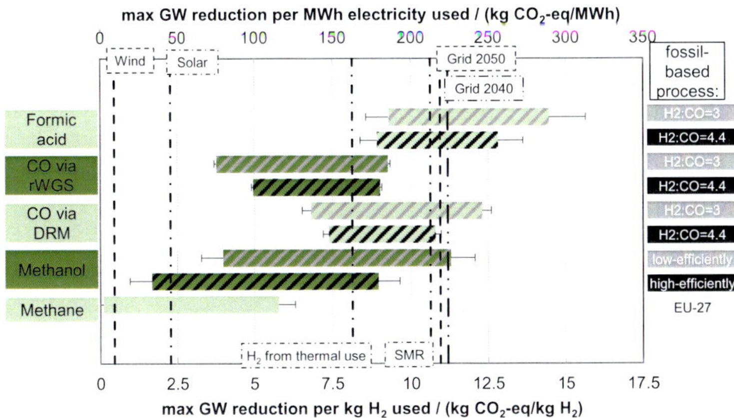

Figure C.6: Maximum global warming (GW) impact reductions for CO_2-based C1-chemicals considering full range for CO_2 supply (GW = -1 to 0 kg CO_2-eq/kg CO_2). For further explanation of the Figure see caption of Figure 4.4 in main text.

Appendix D

Power-to-Gas

D.1 Data for construction of electrolysis and hydrogen storage

Table D.1: Data for construction of electrolysis and hydrogen storage

	Value	Comment
PEM electrolysis	–	inventory is considered equivalent to PEM fuel cell based on rated capacity
life time electrolysis	15 a	value adapted from PEM fuel cell (ecoinvent, 2007b)
life time H_2 storage	50 a	value adapted from liquid storage tank (ecoinvent, 2007a)
steel demand H_2 storage	422.7 t/t H_2	chromium steel 18/8 considered; per hydrogen capacity; (Mori et al., 2014)

D.2 Sensitivity analysis for maximum environmental impact reductions

In this section, a sensitivity analysis is presented for the maximum reduction of global warming and fossil depletion impacts for the aspects summarized in Table D.2.

Table D.2: Aspects considered in sensitivity analysis

		Base case in main text
I)	Utilization of by-products	no environmental credit considered
II)	CO_2 supply (power plant)	Power-to-Gas pathway: power plant with CO_2 capture ($\eta = 27.25$ %) conventional processes: power plant without CO_2 capture ($\eta = 38.27$ %)
III)	Electricity demand of electrolysis	50 kWh per kg H_2
IV)	Methane emissions from natural gas supply	8 g CH_4 per kg natural gas
V)	Operation of electrolysis and	electrolysis: 2,500 full load hours per year
	sizing of H_2 storage	H_2 storage: sized to cover 10 days without renewable electricity supply

I) Utilization of by-products

In contrast to the main text, the environmental credit for the by-products is here based on the avoided burden principle. For this purpose, the avoided burden processes are determined.

Steam. For the steam produced in conventional syngas process, DRM process and SNG process, a credit is given for steam generation from natural gas with an efficiency of 90 %.

Heat. The purge gas produced in the DRM and rWGS process contains CO, hydrogen and methane. We assume thermal utilization of the purge gas. Thus, a credit is given for heat supply by natural gas. The CO_2 emissions of the combustion of the purge gas are determined assuming full combustion, i.e., CO and methane are converted to CO_2.

Oxygen. Oxygen is co-produced in the electrolyzer. The rWGS, DRM and SNG process receive a credit for the production of oxygen by an air separation unit. The environmental impacts of air separation units are dominated by the demand for electricity. For reasons of consistency, we do not use LCA data based on the current grid electricity mix. Rather, the Power-to-Gas routes receive a credit on their electricity demand. The reduced electricity demand of the Power-to-Gas routes increases the maximum reductions of global warming and fossil depletion impacts. The electricity demand of the avoided air separation process is 0.25 kWh / kg O_2 (Hong et al., 2009).

II) CO_2 supply

The scenarios considered in the sensitivity analysis affect either the power plant of the Power-to-Gas pathways or the power plant of the conventional processes. The changes compared to the base case are written in *italics*.

a) Power plant with CO_2 capture without energy penalty. In this case, the coal-fired power plant with CO_2 capture has the same efficiency as the coal-fired power plant without CO_2 capture. This can be considered as the best case for CO_2 supply.

- *Power-to-Gas pathway:* power plant with CO_2 capture *(η = 38.27 %)*
- Conventional processes: power plant without CO_2 capture (η = 38.27 %)

b) Power plant with CO_2 capture vs power plant with CO_2 capture and storage (CCS). In this case, the conventional process also uses a coal-fired power plant with CO_2 capture. The captured CO_2 is stored in the underground. We assume that no additional energy is required for the CO_2 storage.

- Power-to-Gas pathway: power plant with CO_2 capture (η = 27.25 %)
- *Conventional processes:* power plant *with CO_2 capture (η = 27.25 %) and subsequent CO_2 storage in underground*

c) Power plant with CO_2 capture vs renewable electricity. In this case, the conventional process uses a wind power plant to satisfy electricity demand of functional unit.

- Power-to-Gas pathway: coal-fired power plant with CO_2 capture (η = 27.25 %)
- *Conventional processes: wind power plant*

d) Power plant with CO_2 capture supplies electricity to electrolysis. In this case, the electricity of the coal-fired power plant with CO_2 capture is supplied to electrolysis. Thus, the functional unit only contains the production of syngas and SNG, respectively.

- Power-to-Gas pathway: coal-fired power plant with CO_2 capture ($\eta = 27.25$ %)
- *Conventional processes: No power plant required*

III) Electricity demand of electrolysis

a) **45 kWh per kg hydrogen (10 % lower than in main text)**

b) **55 kWh per kg hydrogen (10 % higher than in main text)**

IV) Methane emissions from natural gas supply

In the sensitivity analysis, methane emissions of 17 g per kg natural gas are considered. This value corresponds to the methane emissions for the natural gas supply in Slovakia, the country with the highest methane emissions in Europe according to the database GaBi ts (2016). This scenario is only considered for global warming impacts not for fossil-depletion impacts.

V) Operation of electrolysis and sizing of hydrogen storage

The operation of electrolysis and the sizing of hydrogen storage have an influence on the environmental impacts of construction.

a) **900 full load hours per year for electrolysis**

b) **5,000 full load hours per year for electrolysis**

c) **Hydrogen storage is sized to cover 5 days without renewable electricity supply**

d) **Hydrogen storage is sized to cover 20 days without renewable electricity supply**

D.2.1 Steady-state operation of electrolysis (use of grid electricity)

In this case, the electrolysis is operated in 8,000 h per year and no hydrogen storage is required. Thus, construction of electrolysis and hydrogen storage (IV) is not considered.

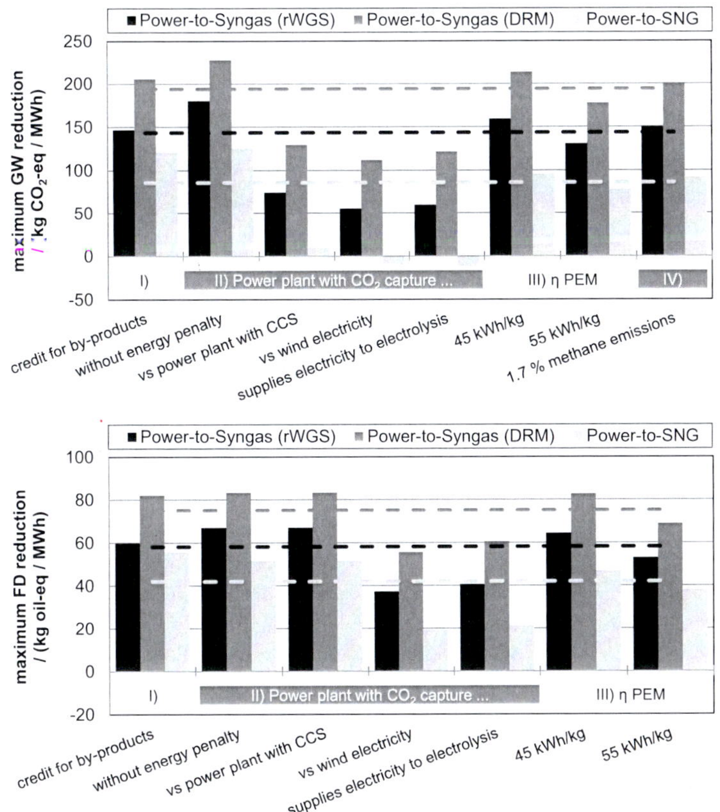

Figure D.1: Sensitivity analysis for maximum impact reductions for global warming (GW, top) and fossil depletion (FD, bottom) per MWh electricity used. The dashed lines indicate the maximum impact reductions for the base case presented in the main text.

D.2.2 Part-load operation of electrolysis (use of 100 % renewable electricity)

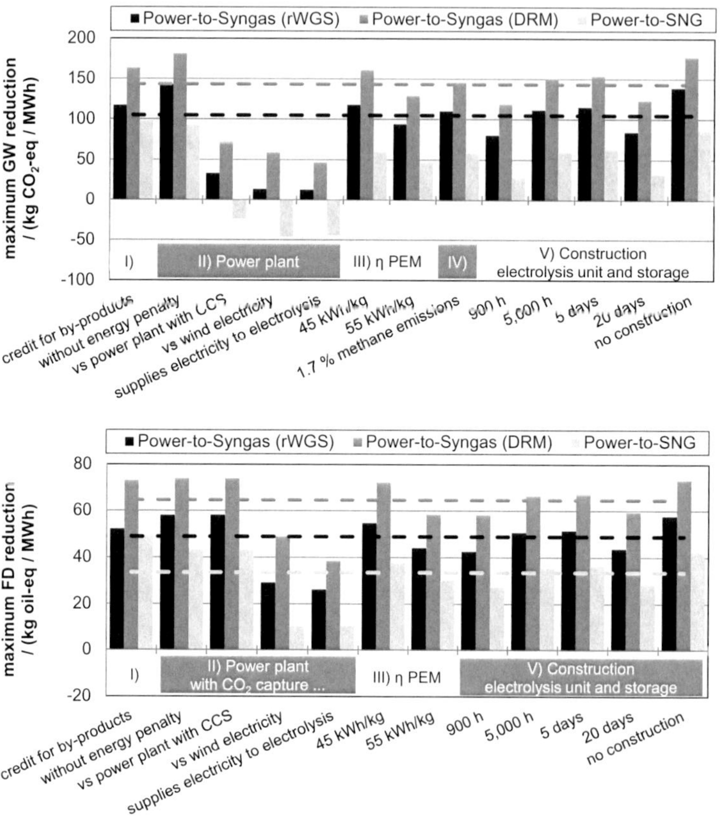

Figure D.2: Sensitivity analysis for maximum impact reductions for global warming (GW, top) and fossil depletion (FD, bottom) per MWh renewable electricity used. The maximum impact reductions refer to a share of 100 % renewable electricity in electrolysis. The dashed lines indicate the maximum impact reductions for the base case presented in the main text.

D.3 Maximum reduction for further environmental impact categories

In this section, maximum reductions are analyzed for 12 further environmental impact categories which are available in ReCiPe. In Table D.3, all environmental impact categories with positive maximum environmental impact reductions are summarized. Only in these environmental impact categories, the Power-to-Gas pathways can achieve lower environmental impacts than the conventional process if an environmentally suitable electricity supply process is chosen. If the forecasted German grid electricity mix for 2050 is applied, lower environmental impacts compared to conventional processes are only achieved for *global warming* and *fossil depletion*.

D.3.1 Steady-state operation of electrolysis

Table D.3: Environmental impact categories with positive maximum reductions

Environmental impact category	rWGS	DRM	SNG
Global warming (in t CO_2-eq/MWh)	1.4E-01	1.9E-01	8.2E-02
Photochemical oxidant formation (in t NMVOC/MWh)	2.3E-06	4.4E-05	–
Fossil depletion (in t oil-eq/MWh)	5.6E-02	7.3E-02	4.0E-02

Comparison of Power-to-Gas pathways to conventional processes

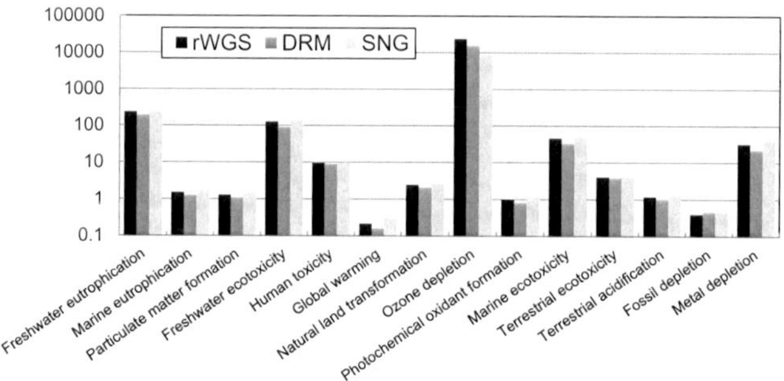

Figure D.3: Environmental impacts (without electricity supply) of Power-to-Gas pathways normalized to corresponding conventional process. If the value for the Power-to-Gas pathways is lower than 1, the corresponding maximum environmental impact reduction is positive.

Hot spots of Power-to-Gas pathways (without electricity supply)

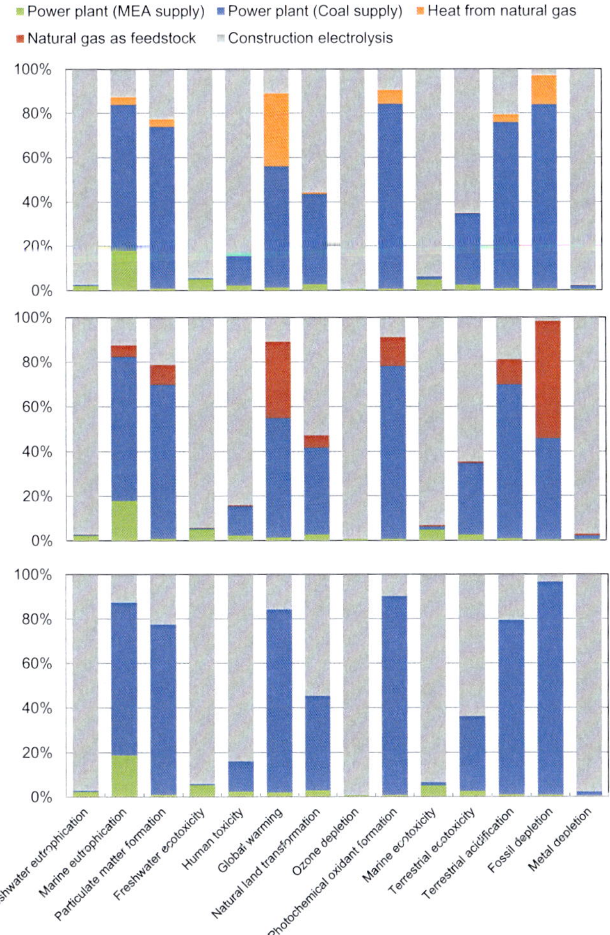

Figure D.4: Environmental hot spots for rWGS (top), DRM (middle) and SNG (bottom) process. Coal supply also includes CO_2 emissions due to combustion of coal in power plant.

D.3.2 Part-load operation of electrolysis

In this case, positive maximum reductions are only achieved in the impact categories global warming and fossil depletion.

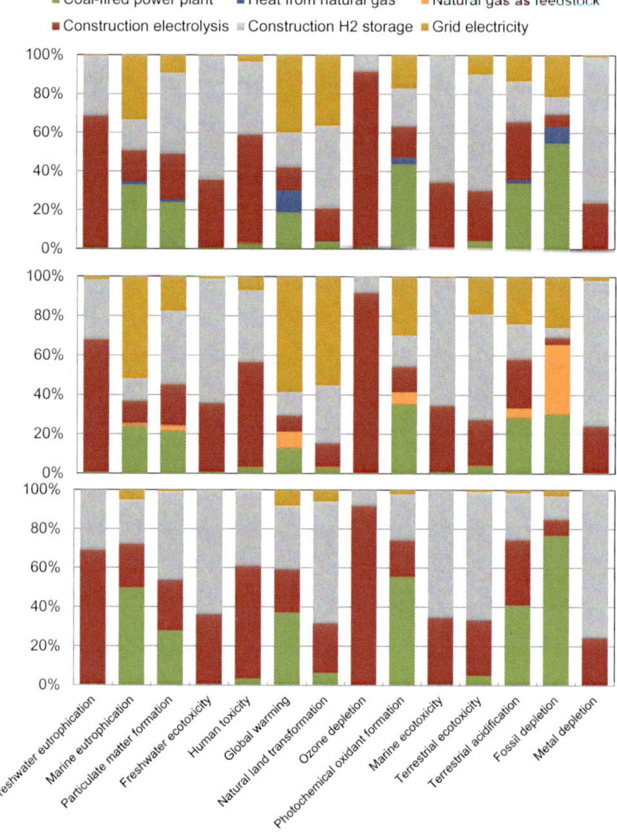

Figure D.5: Environmental hot spots for rWGS (top), DRM (middle) and SNG (bottom) process. Power plant contains coal supply, MEA supply and CO_2 emissions due to combustion of coal in power plant.

Appendix E

Power-to-X

E.1 Surplus electricity supply and demand

Table E.1: Expected surplus electricity in different countries.

Country	Ireland[a]	Australia[b]	
Intermittent power penetration	37 %	88 %	88 %
Wind power penetration	37 %	46 %	59 %
Solar power penetration	–	42 %	29 %
Considered PHS	0.3 GW	2.2 GW	2.2 GW
DSM considered	No	No	No
Surplus electricity	–[c]	9 TWh	25 TWh
share in intermittent power generation	6.6–15.3 %	8.5 %	17.2 %
share in total power generation	2.6–6.7 %	4.1 %	10.9 %
Country		**Germany[d]**	
Intermittent power penetration	84 %	84 %	84 %
Wind power penetration	65 %	65 %	65 %
Solar power penetration	19 %	19 %	19 %
Considered PHS	–	–	8.6 GW
DSM considered	No	Yes	Yes
Surplus electricity	154 TWh	83 TWh	79 TWh
share in intermittent power generation	30.2 %	18.7 %	18.4 %
share in total power generation	22.0 %	13.0 %	12.8 %

[a] Garrigle et al. (2013)
[b] Elliston et al. (2013)
[c] value not mentioned in reference
[d] Klaus et al. (2010)

Table E.2: Renewable electricity demand for global demand of platform chemicals

Platform chemicals	Global demand	Required surplus electricity
Hydrogen	65 Mt/a [a]	3250–5417 TWh/a [d]
Methanol	59 Mt/a [b]	578–964 TWh/a [d,e]
Syngas	2.1 Mt/a [c]	25–41 TWh/a [d,f]

[a] Baufumé et al. (2013)

[b] Jadhav et al. (2014)

[c] syngas demand for polyurethane production (5 Mt/a MDI and 1.9 Mt/a TDI) (Christensen et al., 2013)

[d] for conversion factor Power-to-H_2 see Table E.3 (PEM electrolysis)

[e] for conversion factor H_2-to-Methanol see Figure A.5

[f] for conversion factor H_2-to-Syngas see Figure A.3

E.2 Data for considered Power-to-X systems and conventional processes

Table E.3: Electricity and natural gas demand of renewable power using devices

PHS[a]	$W_{el,out}/W_{el,in}$	0.65–0.85 kWh/kWh
VRB[b]	$W_{el,out}/W_{el,in}$	0.65–0.85 kWh/kWh
CAES[c]	$W_{el,in}/W_{el,out}$	0.7–0.8 kWh/kWh
	$Q_{natural\ gas,in}/W_{el,out}$	1.2–1.3 kWh/kWh
Heat pump[d]	$Q_{heat,out}/W_{el,in}$	2.1–4.9 kWh/kWh
PEM electrolysis[e]	$W_{el,in}/m_{H_2,out}$	45–55 kWh/kg

[a] References: Chatzivasileiadi et al. (2013); Ibrahim et al. (2008) Díaz-González et al. (2012); Ferreira et al. (2013)
[b] References: Chatzivasileiadi et al. (2013); Ibrahim et al. (2008) Ferreira et al. (2013)
[c] Reference: Denholm and Kulcinski (2004)
[d] Reference: Braungardt et al. (2013)
[e] Reference: DOE (2014), case: Future Central Hydrogen Production from PEM Electrolysis version 3.0

Table E.4: Fuel demand of vehicles and fuel cell.

Battery electric vehicle[a]	$W_{el,in}/l_{out}$	0.14–0.20 kWh/km
H_2 fuel cell vehicle[b]	$Q_{H_2,in}/l_{out}$	0.34–0.36 kWh/km
Gasoline engine[c]	$Q_{gasoline,in}/l_{out}$	0.52 kWh/km
Diesel engine[c]	$Q_{diesel,in}/l_{out}$	0.38 kWh/km
Methanol engine[d]	$Q_{CH_3OH,in}/l_{out}$	0.52 kWh/km
Methane engine[e]	$Q_{CH_4,in}/l_{out}$	0.44 kWh/km
Fuel cell[f]	$W_{el,out}/Q_{in}$	0.45–0.60 kWh/kWh

[a] References: Majeau-Bettez et al. (2011); Metz and Doetsch (2012)
[b] References: Offer et al. (2011); Granovskii et al. (2006)
[c] Euro 5, engine displacement $< 1.4\ l$ (GaBi ts, 2016)
[d] gasoline engine efficiency (Ou et al., 2010)
[e] 86.5 % of diesel engine efficiency (Rose et al., 2013)
[f] Reference: Peighambardoust et al. (2010)

Table E.5: Global warming impact (GW) and fossil depletion impact (FD) for construction of Power-to-X systems. All values are per MWh renewable electricity.

Technology	GW kg CO_2-eq	FD kg oil-eq	Life time cycles
Pumped hydro storage[a]	3.2–4.8	1.0–1.4	12,000–25,000
Compresed air energy storage[a]	3.8–5.4	1.3–1.8	8,000–25,000
Vanadium redox flow battery[a]	24.8–58.8	8.2–19.6	2,900–7,500
Lithium-ion battery for battery electric vehicle[b]	36.1–76.7	7.6–16.2	4,000–8,500
Heat pump [c]	76.2–89.2	4.7–7.0	32,000–48,000 h
PEM electrolysis[d]	1.8–2.7	0.7–1.1	
Hydrogen storage[e]	2.2–3.3	0.9–1.3	
Heat storage (hot water tank)[f]	5.1–7.6	1.7–2.5	20–30 a
Methanol storage[f]	2.4–3.7E-02	7.8–12E-03	40–60 a
Methane plant[g]	1.8–1.8E-02	4.9–7.3E-03	
Methanol plant[h]	6.9–10E-02	2.0–3.0E-02	24–36 a
Syngas plant[h]	6.4–9.6E-02	1.8–2.6E-02	40–60 a
PEM fuel cell[i]	1.8–2.7	0.7–1.1	32,000–48,000 h

[a] References: Denholm and Kulcinski (2004); Barnhart and Benson (2013)
[b] References: Majeau-Bettez et al. (2011); Barnhart and Benson (2013)
[c] Reference: ecoinvent Data V 2.2 (2010); values include leakage of refrigerant
[d] value of fuel cell assumed (Manish et al., 2006)
[e] Reference: Spath and Mann (2004); life time not mentioned in reference
[f] Reference: ecoinvent Data V 2.2 (2010); utilization: 200 cycles/a
[g] own calculation based on syngas plant
[h] Reference: ecoinvent Data V 2.2 (2010)
[i] Reference: Pehnt (2001); 90 % platinum group metal recycling assumed

Table E.6: Costs for Power-to-Power, Power-to-Mobility, Power-to-Heat, Power-to-H_2 storage systems and conventional systems. All cost are for 2020.

Technology	Power capacity capital costs	Energy capacity capital costs	variable O&M costs	fixed O&M costs	life time
	$/kW	$/kWh	$/kWh	$/(kW·a)	a
Pumped hydro storage[a]	908–1236	12.3–17.4	-	3.0	40
Compressed air energy storage[a]	412–501	1.2–62.8	-	3.0–7.1	40
Vanadium redox flow battery[a]	537–1840	215–307	-	56.0	40
Natural gas turbine[b]	660	-	0.03	5.3	40
Electric drive train for BEV[c]	1200–2030 $	-	-	-	13.2
Battery for BEV[d]	-	273–386	-	-	13.2
Conventional drive train[c]	2400–2530 $	-	-	-	13.2
Heat pump[e]	750–1147	-	-	5	20
Electric boiler[e]	140–220	-	-	-	20
Heat storage[f]	-	90–170	-	-	40
Natural gas boiler[e]	725	-	-	4.4	20
PEM electrolysis[g]	465–698	-	-	35.4	40
Hydrogen storage[h]	-	390–584	-	-	30

O&M - operation and maintenance

[a] Reference: Steward et al. (2009)

[b] Reference: Black & Veatch (2012)

[c] Reference: Offer et al. (2011)

[d] Reference: EPA (2012)

[e] Reference: Nitsch et al. (2010); in Euro

[f] Reference: Hedegaard et al. (2012); in Euro

[g] Reference: DOE (2014)

[h] Reference: DOE (2014); in $/kg H_2

For the determination of CO_2 mitigation cost in Euro, an exchange rate of 1 Euro = 1.3 US Dollar is assumed.

Table E.7: Costs for feedstocks and conventional produced hydrogen for 2020 in the US

Feedstock	Costs	
Residential natural gas	49.6 \$/ MWh	(DOE, 2014)
Industrial natural gas	29.5 \$/ MWh	(DOE, 2014)
Industrial electricity	67.1 \$/ MWh	(DOE, 2014)
Gasoline	102.6 \$/ MWh	(Offer et al., 2011)
CO_2	70 \$/ t	(Plasynski and Ciferno, 2008)
Hydrogen	1.6 \$/ kg	(DOE, 2014)

Table E.8: Inputs per kg product for conventional production of methanol and syngas

Product	natural gas	electricity	
Methanol	9.03 kWh	0.074 kWh	(ecoinvent Data V 2.2, 2010)
Syngas	10.44 kWh	0.713 kWh	(CO_2RRECT, 2014)

E.3 Results

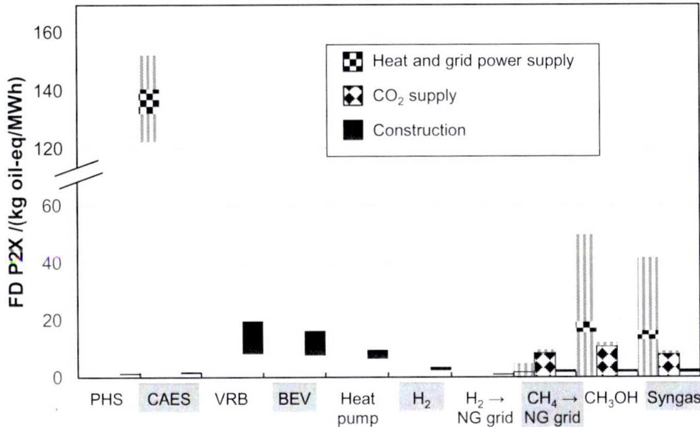

Figure E.1: Breakdown of fossil depletion (FD) impact of Power-to-X systems into considered processes. All values are per MWh renewable electricity. The narrow ranges (patterned) represent average LCA data sets for the EU-27 considering the technology uncertainties. The broad ranges (gray striped) further include the full range of LCA data sets for all countries. If only one range is shown, fossil depletion impacts are independent from country-specific LCA data sets. For CO_2 supply, only the fossil depletion impact for the scenario *CO_2 emissions avoided* is shown. For the scenario *CO_2 storage avoided*, the corresponding fossil depletion impact would be zero. PHS - pumped hydro storage; CAES - compressed air energy storage; VRB - vanadium redox flow battery; BEV - battery electric vehicle; NG - natural gas

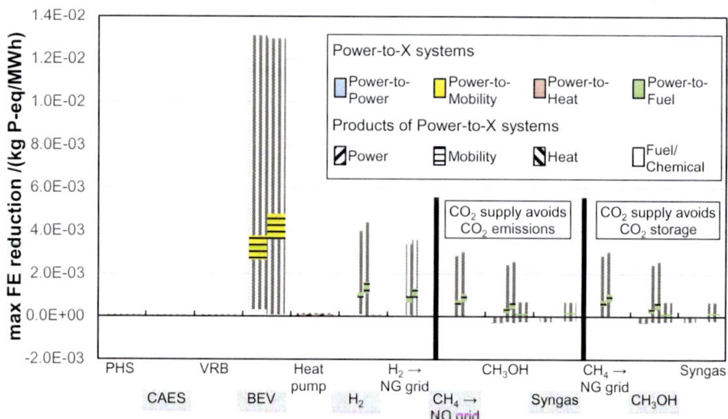

Figure E.2: Freshwater eutrophication (FE) impact reduction for Power-to-X systems. The narrow ranges (colored) represent average LCA data sets for the EU-27 considering the technology uncertainties. The broad ranges (gray striped) further include the full range of LCA data sets for all countries. For Power-to-Fuel systems, impact reductions are shown for the different utilization routes of the products. For CO_2-using Power-to-Fuel systems, the freshwater eutrophication impact reductions are shown for both CO_2 supply scenarios (*CO_2 emissions avoided* and *CO_2 storage avoided*). PHS - pumped hydro storage; CAES - compressed air energy storage; VRB - vanadium redox flow battery; BEV - battery electric vehicle; NG - natural gas

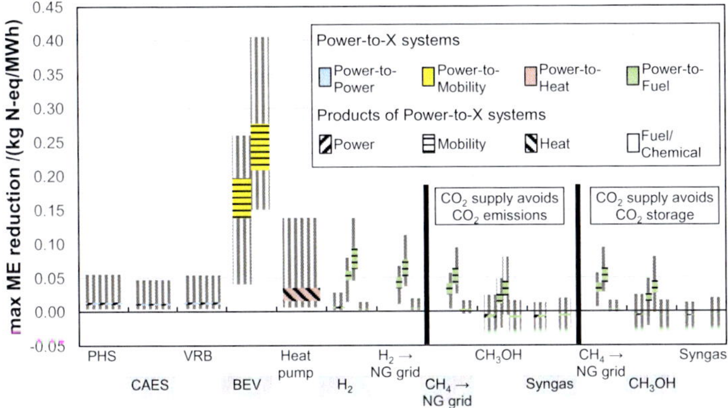

Figure E.3: Marine eutrophication (ME) impact reduction for Power-to-X systems. For further information see caption of Figure E.2.

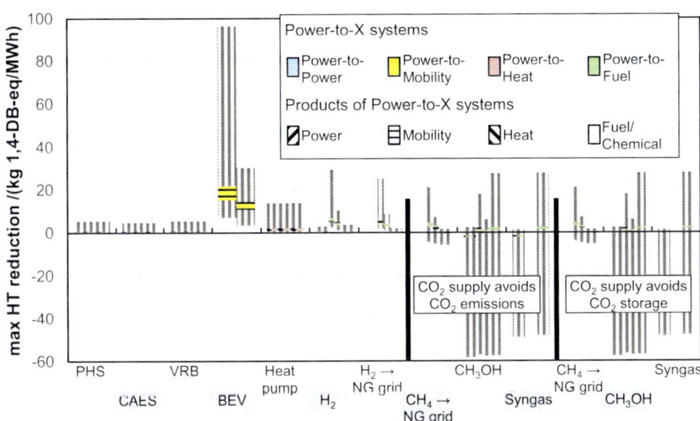

Figure E.4: Human toxicity (HT) impact reduction for Power-to-X systems. For further information see caption of Figure E.2.

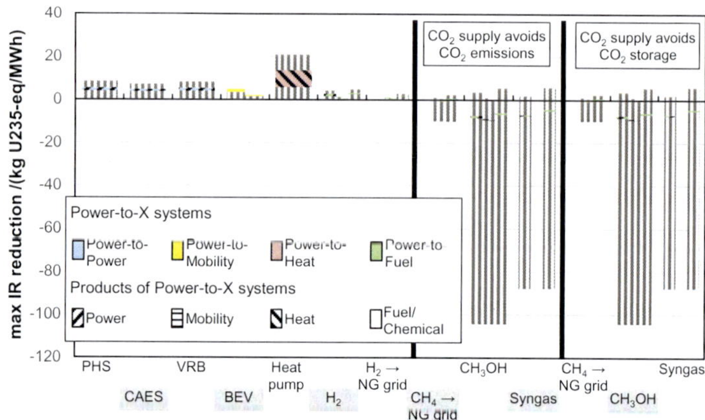

Figure E.5: Ionizing radiation (IR) impact reduction for Power-to-X systems. For further information see caption of Figure E.2.

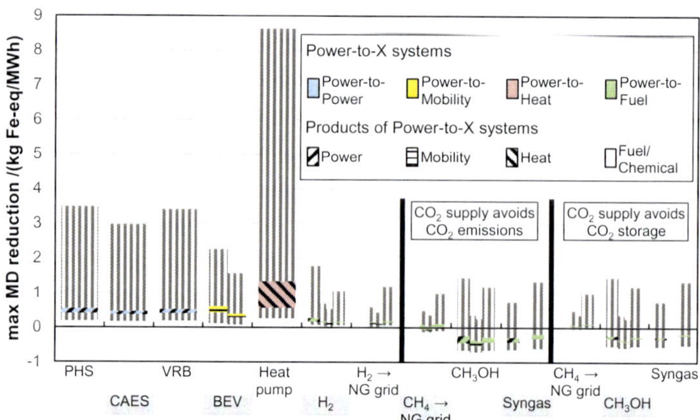

Figure E.6: Mineral resource depletion (MD) impact reduction for Power-to-X systems. For further information see caption of Figure E.2.

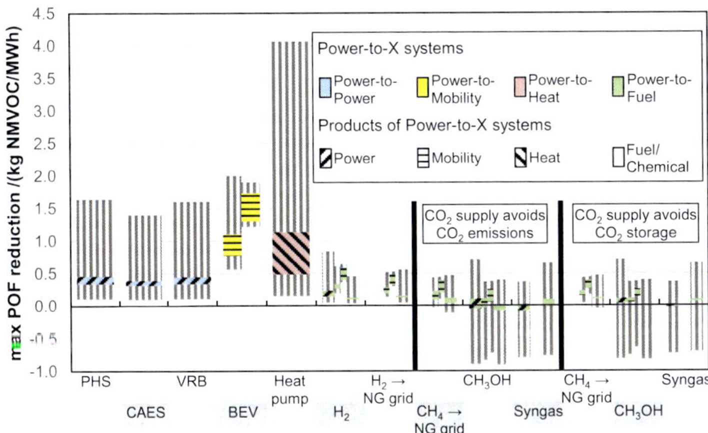

Figure E.7: Photochemical oxidant formation (POF) impact reduction for Power-to-X systems. For further information see caption of Figure E.2.

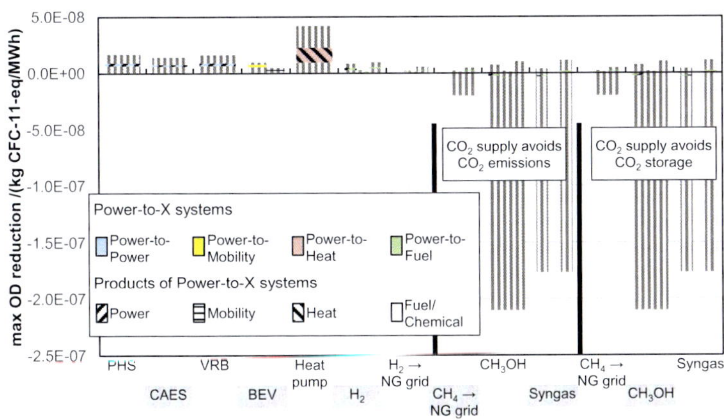

Figure E.8: Ozone depletion (OD) impact reduction for Power-to-X systems. For further information see caption of Figure E.2.

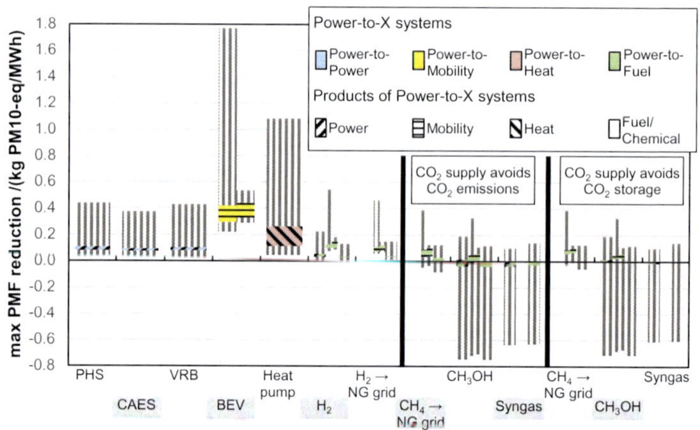

Figure E.9: Particulate matter formation (PMF) impact reduction for Power-to-X systems. For further information see caption of Figure E.2.

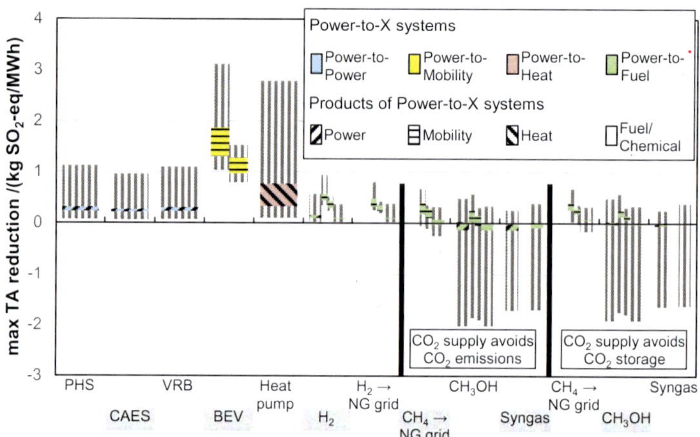

Figure E.10: Terrestrial acidification (TA) impact reduction for Power-to-X systems. For further information see caption of Figure E.2.

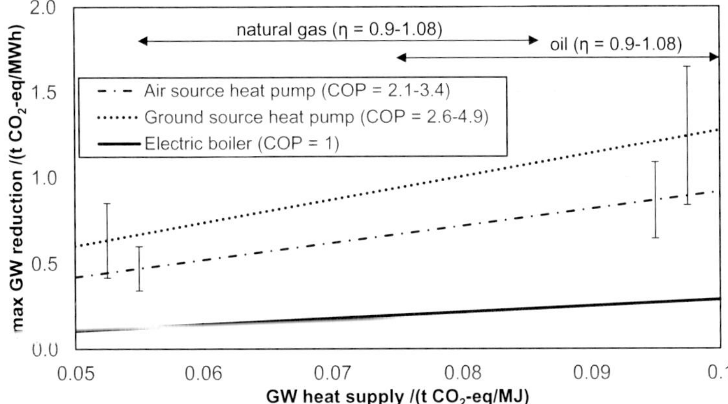

Figure E.11: Maximum global warming impact reductions for Power-to-Heat systems depending on the global warming impact of the conventional process (GW heat supply). The error bars indicate the possible range for each Power-to-Heat system considering the presented coefficient of performance (COP). On top of the graph, the typical range for the conventional processes heat from oil and heat from natural gas is shown. This range includes the presented range of efficiencies and LCA data sets for different countries.

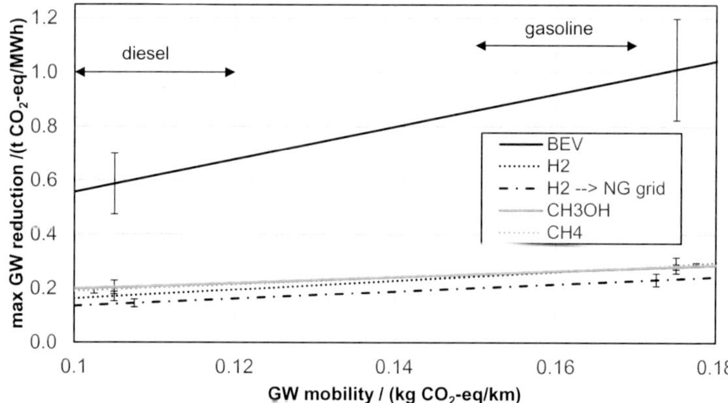

Figure E.12: Maximum global warming impact reductions for Power-to-Mobility systems depending on the global warming impact of the conventional process (GW mobility). The error bars indicate the possible range for each Power-to-Mobility system considering the presented efficiencies in Tables E.3 and E.4. On top of the graph, the typical range for the conventional processes mobility from gasoline and mobility from diesel is shown. This range includes LCA data sets for different countries.

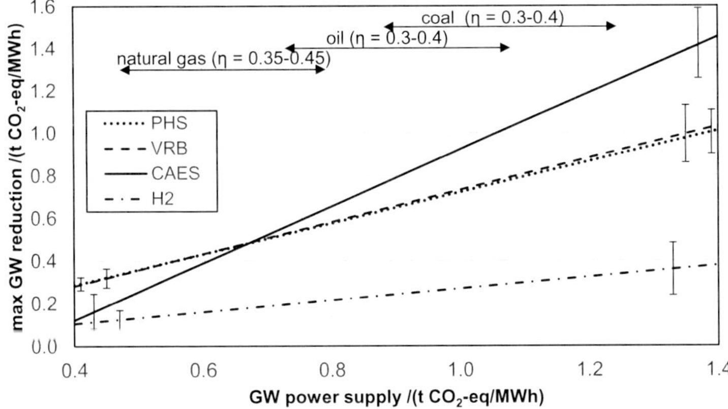

Figure E.13: Maximum global warming impact reductions for Power-to-Power systems depending on the global warming impact of the conventional process (GW power supply). The error bars indicate the possible range for each Power-to-Power system considering the presented efficiencies in Tables E.3 and E.4. On top of the graph, the typical range for the conventional processes power from natural gas, oil and coal is shown. This range includes the presented range of efficiencies and LCA data sets for different countries.

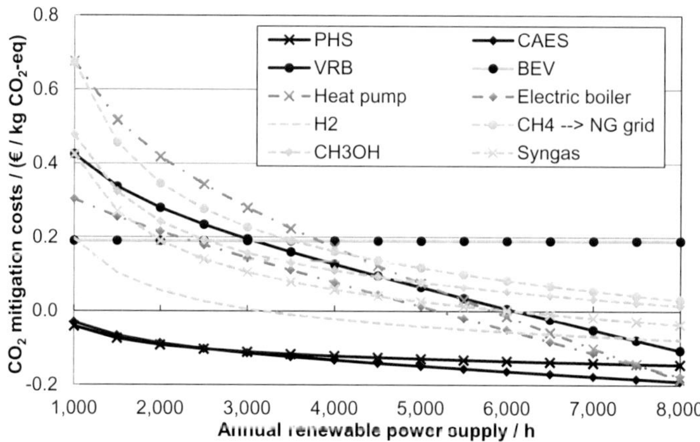

Figure E.14: CO_2 mitigation costs of Power-to-X systems as a function of available renewable power. In this case, renewable power occurs once a week.

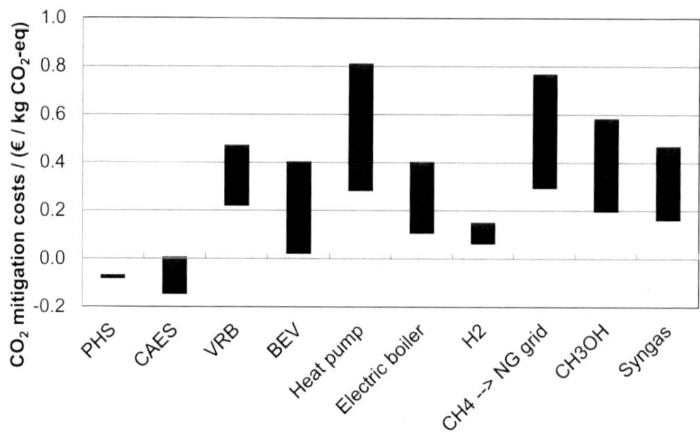

Figure E.15: CO_2 mitigation costs of Power-to-X systems for 1,500 hours with renewable power supply. In this case, renewable power occurs once a week. The ranges represent US-specific LCA data sets considering the technology uncertainty and the economic uncertainty presented in Table E.6.

Appendix F

Student theses completed within this work

Budke, J. (2013) Bewertung des Einflusses von Demand Side Management auf den Kraftwerkseinsatz in Europa (in German). Master thesis, RWTH Aachen University.

Deric, S. (2015). Prozesssimulation zur ökologischen Bewertung von C1-Chemikalien aus Kohlendioxid (in German). Bachelor thesis, RWTH Aachen University.

Hausfeld, T. (2012). Entwicklung eines Biogasanlagenkonzepts zur Direktvermarktung auf Basis einer technischen und ökonomischen Analyse (in German). Diploma thesis. RWTH Aachen University.

Henke, F. (2013). Optimierung des Einsatzes von Pumpspeicherkraftwerken auf Spot- und Regelenergiemärkten unter Berücksichtigung von Preisunsicherheiten (in German). Master thesis, RWTH Aachen University.

Hewing, J. (2012). Untersuchung von Stoffmodellen für Wasser zur Simulation von Energiesystemen (in German). Bachelor thesis, RWTH Aachen University.

Lohaus, M. (2013). Optimierung von Wasserstoffspeichersystemen unter Berücksichtigung von Unsicherheiten (in German). Master thesis, RWTH Aachen University.

Lühn, T. (2012). Ökonomische und ökologische Bewertung einer Power-to-Gas Anlage zur Integration erneuerbarer Energien in die vorhandene Netzinfrastruktur der Region Münsingen (in German). Diploma thesis. RWTH Aachen University.

Teichgräber, H. (2014). Strukturoptimierung zur Identifikation ökologischer Prozessrouten für C1-Chemikalien aus Biomasse und Kohlenstoffdioxid (in German). Bachelor thesis, RWTH Aachen University.

Teichgräber, H. (2013). A Simulation Model for Sizing of Hybrid Renewable Energy Systems for Residential Applications. Project thesis, University of California Davis.

von Wolff, S. and Petersdorff-Campen, C. (2013). Zeitlich aufgelöste Bestimmung der Treibhausgasemissionen der deutschen Stromerzeugung (in German). Project thesis, RWTH Aachen University.

Wurm, A. and Oltmanns, J. G. (2011). Konzeptionierung und techno-ökonomische Bewertung energieeffizienter Kuhlvarianten für den Schaltschrank von Werkzeugmaschinen (in German). Project thesis, RWTH Aachen University.

Bibliography

Ampelli, C., Perathoner, S., and Centi, G. (2015). CO_2 utilization: An enabling element to move to a resource- and energy-efficient chemical and fuel production. *Philos. T. Roy. Soc. A*, 373(2037):20140177.

Anderson, J. J., Drury, D. J., Hamlin, J. E., and Kent, A. G. (1986). Patent WO 8602066.

Aresta, M., Caroppo, A., Dibenedetto, A., and Narracci, M. (2002). Life cycle assessment (LCA) applied to the synthesis of methanol. Comparison of the use of syngas with the use of CO_2 and dihydrogen produced from renewables. In Maroto-Valer, M., Song, C., and Soong, Y., editors, *Environmental Challenges and Greenhouse Gas Control for Fossil Fuel Utilization in the 21st Century*, pages 331–347. Springer US.

Aresta, M., Dibenedetto, A., and Angelini, A. (2013). The changing paradigm in CO_2 utilization. *J. CO_2 Util.*, 3-4:65–73.

Asinger, F. (1986). *Methanol: Chemie- und Energierohstoff*. Springer, Berlin Heidelberg New York.

Baltrusaitis, J. and Luyben, W. L. (2015). Methane conversion to syngas for Gas-to-Liquids (GTL): Is sustainable CO_2 reuse via dry methane reforming (DMR) cost competitive with SMR and ATR processes? *ACS Sustainable Chem. Eng.*, 3(9):2100–2111.

Barnhart, C. J. and Benson, S. M. (2013). On the importance of reducing the energetic and material demands of electrical energy storage. *Energy Environ. Sci.*, 6(4):1083–1092.

Baufumé, S., Grüger, F., Grube, T., Krieg, D., Linssen, J., Weber, M., Hake, J.-F., and Stolten, D. (2013). GIS-based scenario calculations for a nationwide German hydrogen pipeline infrastructure. *Int. J. Hydrogen Energy*, 38(10):3813–3829.

Beaudin, M., Zareipour, H., Schellenberglabe, A., and Rosehart, W. (2010). Energy storage for mitigating the variability of renewable electricity sources: An updated review. *Energy Sust. Dev.*, 14(4):302–314.

Bertuccioli, L., Chan, A., Hart, D., Lehner, F., Madden, B., and Standen, E. (2014). Study on development of water electrolysis in the EU. Final report, E4tech Sarl with Element Energy Ltd.

Black & Veatch (2012). Cost and perfromance data for power generation technologies. Cost Report for NREL, Black & Veatch Holding Company, Overland Park, KS.

Bontemps, S., Vendier, L., and Sabo-Etienne, S. (2014). Ruthenium-catalyzed reduction of carbon dioxide to formaldehyde. *J. Am. Chem. Soc.*, 136(11):4419–4425.

Braungardt, S., Günther, D., Miara, M., Wapler, J., and Weßing, W. (2013). *Electrically driven heat pumps.* FIZ Karlsruhe GmbH, Karlsruhe.

Centi, G., Quadrelli, E. A., and Perathoner, S. (2013). Catalysis for CO_2 conversion: A key technology for rapid introduction of renewable energy in the value chain of chemical industries. *Energy Environ. Sci.*, 6(6):1711–1731.

Chatzivasileiadi, A., Ampatzi, E., and Knight, I. (2013). Characteristics of electrical energy storage technologies and their applications in buildings. *Renew. Sust. Energy Rev.*, 25:814–830.

Chen, L., Lu, Y., Hong, Q., Lin, J., and Dautzenberg, F. (2005). Catalytic partial oxidation of methane to syngas over Ca-decorated-Al_2O_3-supported Ni and NiB catalysts. *Appl. Catal. A-Gen.*, 292:295–304.

Christensen, F., Nilsson, N. H., Jeppesen, C. N., and Clausen, A. J. (2013). Survey of certain isocyanates (MDI and TDI). Environmental Project No. 1537, The Danish Environmental Protection Agency, Copenhagen, DK.

Clausen, L. R., Houbak, N., and Elmegaard, B. (2010). Technoeconomic analysis of a methanol plant based on gasification of biomass and electrolysis of water. *Energy*, 35(5):2338–2347.

CO_2RRECT (2014). CO_2-Reaction using Regenerative Energies and Catalytic Technologies. Final project report, (ref. no. 033RC1006B).

Denholm, P. and Kulcinski, G. L. (2004). Life cycle energy requirements and greenhouse gas emissions from large scale energy storage systems. *Energy Convers. Manage.*, 45(13-14):2153–2172.

Díaz-González, F., Sumper, A., Gomis-Bellmunt, O., and Villafáfila-Robles, R. (2012). A review of energy storage technologies for wind power applications. *Renew. Sust. Energy Rev.*, 16(4):2154–2171.

DOE (2014). H2a production analysis. Technical report, US Department of Energy, Washington, DC.

Dumont, M.-N., von der Assen, N., Sternberg, A., and Bardow, A. (2012). Assessing the environmental potential of carbon dioxide utilization: A graphical targeting approach. In Karimi, I. A. and Srinivasan, R., editors, *11th International Symposium on Process Systems Engineering*, volume 31 of *Computer Aided Chemical Engineering*, pages 1407–1411. Elsevier.

ecoinvent (2007a). Life cycle inventories of chemicals. ecoinvent report No. 8, Swiss Centre for Life Cycle Inventories, Zurich, CH.

ecoinvent (2007b). Life cycle inventories of new CHP systems. ecoinvent report No. 20, Swiss Centre for Life Cycle Inventories, Zurich, CH.

ecoinvent (2007c). Life cycle inventories of petrochemical solvents. ecoinvent report No. 22, Swiss Centre for Life Cycle Inventories, Zurich, CH.

ecoinvent Data V 2.2 (2010). Swiss Centre for Life Cycle Inventories. (accessed August 2016).

Economides, M. J. and Wood, D. A. (2009). The state of natural gas. *J. Nat. Gas Sci. Eng.*, 1(1–2):1–13.

Ekvall, T. and Weidema, B. (2004). System boundaries and input data in consequential life cycle inventory analysis. *Int. J. Life Cycle Assess.*, 9(3):161–171.

Elliston, B., MacGill, I., and Diesendorf, M. (2013). Least cost 100 % renewable electricity scenarios in the Australian national electricity market. *Energy Policy*, 59:270–282.

EPA (2012). Modeling the cost and performance of lithium-ion batteries for electric-drive vehicles. EPA-420-R-12-023, US Environmental Protection Agency, Washington, DC.

Ferreira, H. L., Garde, R., Fulli, G., Kling, W., and Lopes, J. P. (2013). Characterisation of electrical energy storage technologies. *Energy*, 53:288–298.

Fiedler, E., Grossmann, G., Kersebohm, D. B., Weiss, G., and Witte, C. (2000). *Ullmann's Encyclopedia of Industrial Chemistry*, chapter Methanol. Wiley-VCH Verlag GmbH & Co. KGaA.

Finnveden, G., Hauschild, M. Z., Ekvall, T., Guinée, J., Heijungs, R., Hellweg, S., Koehler, A., Pennington, D., and Suh, S. (2009). Recent developments in life cycle assessment. *J. Environ. Manage.*, 91(1):1–21.

Foley, A. and Lobera, I. D. (2013). Impacts of compressed air energy storage plant on an electricity market with a large renewable energy portfolio. *Energy*, 57:85–94.

GaBi ts (2016). Software-System and Database for Life Cycle Engineering. thinkstep AG, Leinfelden-Echterdingen, DE.

Garrigle, E. M., Deane, J., and Leahy, P. (2013). How much wind energy will be curtailed on the 2020 Irish power system? *Renewable Energy*, 55:544–553.

Gassner, M. and Maréchal, F. (2008). Thermo-economic optimisation of the integration of electrolysis in synthetic natural gas production from wood. *Energy*, 33(2):189–198.

Gassner, M. and Maréchal, F. (2012). Thermo-economic optimisation of the polygeneration of synthetic natural gas (SNG), power and heat from lignocellulosic biomass by gasification and methanation. *Energy Environ. Sci.*, 5(2):5768–5789.

Gerdes, K., Haslbeck, J., Kuehn, N., Lewis, E., Pinkerton, L. L., Woods, M., Simpson, J., Turner, M. J., and Varghese, E. (2013). Cost and performance baseline for fossil energy plants - volume 1: Bituminous coal and natural gas to electricity (revision 2a). Report DOE/NETL-2010/1397, National Energy Technology Laboratory, Washington, DC.

Goedkoop, M., Heijungs, R., Huijbregts, M., Schryver, A. D., Struijs, J., and Zelm, R. V. (2009). *ReCiPe 2008, A life cycle impact assessment method which comprises harmonised category indicators at the midpoint and the endpoint level; First edition Report I: Characterisation.*

Götz, M., Lefebvre, J., Mörs, F., Koch, A. M., Graf, F., Bajohr, S., Reimert, R., and Kolb, T. (2016). Renewable power-to-gas: A technological and economic review. *Renewable Energy*, 85:1371–1390.

Granovskii, M., Dincer, I., and Rosen, M. A. (2006). Economic and environmental comparison of conventional, hybrid, electric and hydrogen fuel cell vehicles. *J. Power Sources*, 159(2):1186–1193.

Green, M. J., Lucy, A. R., Kitson, M., and Smith, S. J. (1989). Patent EP 0329337.

Gu, Y., Xu, J., Wang, H., and Li, F. (2014). Industrial water footprint assessment: Methodologies in need of improvement. *Environ. Sci. Technol.*, 48(12):6531–6532.

Hamelinck, C. N. and Faaij, A. P. (2002). Future prospects for production of methanol and hydrogen from biomass. *J. Power Sources*, 111(1):1–22.

Hammer, G., Lübcke, T., Kettner, R., Davis, R. N., Recknagel, H., Commichau, A., Neumann, H.-J., and Paczynska-Lahme, B. (2000). *Ullmann's Encyclopedia of Industrial Chemistry*, chapter Natual gas. Wiley-VCH Verlag GmbH & Co. KGaA.

Hauschild, M., Goedkoop, M., Guinée, J., Heijungs, R., Huijbregts, M., Jolliet, O., Margni, M., De Schryver, A., Humbert, S., Laurent, A., Sala, S., and Pant, R. (2013). Identifying best existing practice for characterization modeling in life cycle impact assessment. *Int. J. Life Cycle Assess.*, 18(3):683–697.

Hedegaard, K., Mathiesen, B. V., Lund, H., and Heiselberg, P. (2012). Wind power integration using individual heat pumps - analysis of different heat storage options. *Energy*, 47(1):284–293.

Hennings, W., Mischinger, S., and Linssen, J. (2013). Utilization of excess wind power in electric vehicles. *Energy Policy*, 62:139–144.

Henriksson, P., Guinée, J., Heijungs, R., Koning, A., and Green, D. (2014). A protocol for horizontal averaging of unit process data-including estimates for uncertainty. *Int. J. Life Cycle Assess.*, 19(2):429–436.

Hewitt, N. J. (2012). Heat pumps and energy storage - the challenges of implementation. *Appl. Energy*, 89(1):37–44.

Hietala, J., Vuori, A., Johnsson, P., Pollari, I., Reutemann, W., and Kieczka, H. (2016). *Ullmann's Encyclopedia of Industrial Chemistry*, chapter Formic Acid. Wiley-VCH Verlag GmbH & Co. KGaA.

Hong, J., Chaudhry, G., Brisson, J., Field, R., Gazzino, M., and Ghoniem, A. F. (2009). Analysis of oxy-fuel combustion power cycle utilizing a pressurized coal combustor. *Energy*, 34(9):1332–1340.

Ibrahim, H., Ilinca, A., and Perron, J. (2008). Energy storage systems - characteristics and comparisons. *Renew. Sust. Energy Rev.*, 12(5):1221–1250.

IEA (2016). IEA Sankey Diagram. http://www.iea.org/Sankey/. (accessed April 21, 2016).

ISO 14040 (2009). Environmental management – Life cycle assessment – Principles and framework. (ISO 14040:2006), European Committee for Standardisation, Brussels, BE.

ISO 14044 (2006). Environmental management – Life cycle assessment – Requirements and guidelines. (ISO 14044:2006), European Committee for Standardisation, Brussels, BE.

Jadhav, S. G., Vaidya, P. D., Bhanage, B. M., and Joshi, J. B. (2014). Catalytic carbon dioxide hydrogenation to methanol: A review of recent studies. *Chem. Eng. Res. Des.*, 92(11):2557–2567.

Jafarbegloo, M., Tarlani, A., Mesbah, A. W., and Sahebdelfar, S. (2015). Thermodynamic analysis of carbon dioxide reforming of methane and its practical relevance. *Int. J. Hydrogen Energy*, 40(6):2445–2451.

Jean, M. D. S., Baurens, P., and Bouallou, C. (2014). Parametric study of an efficient renewable power-to-substitute-natural-gas process including high-temperature steam electrolysis. *Int. J. Hydrogen Energy*, 39(30):17024–17039.

Jens, C. M., Nowakowski, K., Scheffczyk, J., Leonhard, K., and Bardow, A. (2016). CO from CO_2 and fluctuating renewable energy via formic-acid derivatives. *Green Chem.*, DOI: 10.1039/C6GC01202G.

Jens, C. M., Scott, M., Liebergesell, B., Schäfer, P., Franciò, G., Leitner, W., Leonhard, K., and Bardow, A. (2017). Balancing reaction and separation in direct CO_2-based formic acid synthesis with dimethyl sulfoxide/co-solvent mixtures. *in preparation*.

Jiang, Y., Blacque, O., Fox, T., and Berke, H. (2013). Catalytic CO_2 activation assisted by rhenium hydride/$B(C_6F_5)_3$ frustrated Lewis pairs–metal hydrides functioning as FLP bases. *J. Am. Chem. Soc.*, 135(20):7751–7760.

Jung, J., Postels, S., and Bardow, A. (2014). Cleaner chlorine production using oxygen depolarized cathodes? A life cycle assessment. *J. Clean. Prod.*, 80:46–56.

Jung, J., von der Assen, N., and Bardow, A. (2013). Comparative LCA of multi-product processes with non-common products: A systematic approach applied to chlorine electrolysis technologies. *Int. J. Life Cycle Assess.*, 18(4):828–839.

Kaiser, P., Unde, R., Kern, C., and Jess, A. (2013). Production of liquid hydrocarbons with CO_2 as carbon source based on reverse water-gas shift and Fischer-Tropsch synthesis. *Chem. Ing. Tech.*, 85(4):489–499.

Kaltschmitt, M., Streicher, W., and Wiese, A., editors (2013). *Erneuerbare Energien - Systemtechnik, Wirtschaftlichkeit, Umweltaspekte.* Springer-Verlag Berlin Heidelberg, 5 edition. (in German).

Kätelhön, A., von der Assen, N., Suh, S., Jung, J., and Bardow, A. (2015). Industry-cost-curve approach for modeling the environmental impact of introducing new technologies in life cycle assessment. *Environ. Sci. Technol.*, 49(13):7543–7551.

Kiss, A. A., Pragt, J., Vos, H., Bargeman, G., and de Groot, M. (2016). Novel efficient process for methanol synthesis by CO_2 hydrogenation. *Chem. Eng. J.*, 284:260–269.

Klankermayer, J., Wesselbaum, S., Beydoun, K., and Leitner, W. (2016). Selective catalytic synthesis using the combination of carbon dioxide and hydrogen. *Angew. Chem. Int. Ed.*, 55(26):7296–7343.

Klaus, T., Vollmer, C., Werner, K., Lehmann, H., and Müschenüschen, K. (2010). *Energy target 2050.* Study UBA - FKZ 363 01 277.

Klöpffer, W. and Grahl, B. (2009). *Ökobilanz LCA Ein Leitfaden für Ausbildung und Beruf.* Wiley-VCH Verlag GmbH & Co. KGaA.

Kondratenko, E. V., Mul, G., Baltrusaitis, J., Larrazabal, G. O., and Perez-Ramirez, J. (2013). Status and perspectives of CO_2 conversion into fuels and chemicals by catalytic, photocatalytic and electrocatalytic processes. *Energy Environ. Sci.*, 6(11):3112–3135.

Kreimeyer, A. (2013). New directions in industrial chemical research as reflected in Angewandte Chemie. *Angew. Chem. Int. Ed.*, 52(1):147–154.

Ledon, H. (1986). *Ullmann's Encyclopedia of Industrial Chemistry*, chapter Carbon monoxide. Wiley-VCH Verlag GmbH & Co. KGaA.

Leitner, W. (1995). Carbon dioxide as a raw material: The synthesis of formic acid and its derivatives from CO_2. *Angew. Chem. Int. Ed.*, 34(20):2207–2221.

Majeau-Bettez, G., Hawkins, T. R., and Strømman, A. H. (2011). Life cycle environmental assessment of lithium-ion and nickel metal hydride batteries for plug-in hybrid and battery electric vehicles. *Environ. Sci. Technol.*, 45(10):4548–4554.

Manish, S., Pillai, I. R., and Banerjee, R. (2006). Sustainability analysis of renewables for climate change mitigation. *Energy Sust. Dev.*, 10(4):25–36.

Metz, M. and Doetsch, C. (2012). Electric vehicles as flexible loads - a simulation approach using empirical mobility data. *Energy*, 48(1):369–374.

Moret, S., Dyson, P. J., and Laurenczy, G. (2014). Direct synthesis of formic acid from carbon dioxide by hydrogenation in acidic media. *Nat Commun*, 5:4017.

Mori, M., Jensterle, M., Mržljak, T., and Drobnič, B. (2014). Life-cycle assessment of a hydrogen-based uninterruptible power supply system using renewable energy. *Int. J. Life Cycle Assess.*, 19(11):1810–1822.

Moriarty, P. and Honnery, D. (2016). Can renewable energy power the future? *Energy Policy*, 93:3–7.

Müller, B., Müller, K., Teichmann, D., and Arlt, W. (2011). Energiespeicherung mittels Methan und energietragenden Stoffen - ein thermodynamischer Vergleich. *Chem. Ing. Tech.*, 83(11):2002–2013 (in German).

Müller, K., Geng, J., Völkl, J., and Arlt, W. (2012). Energetische Betrachtung der Wasserstoffeinspeisung ins Erdgasnetz. *Chem. Ing. Tech.*, 84(9):1513–1519 (in German).

Nitsch, J., Pregger, T., Scholz, Y., Naegler, T., Sterner, M., Gerhardt, N., von Oehsen, A., Pape, C., Saint-Drenan, Y., and Wenzel, B. (2010). *Langfristszenarien und Strategien für den Ausbau der erneuerbaren Energien in Deutschland bei Berücksichtigung der Entwicklung in Europa und global.* Study BMU - FKZ 03MAP146.

Offer, G., Contestabile, M., Howey, D., Clague, R., and Brandon, N. (2011). Techno-economic and behavioural analysis of battery electric, hydrogen fuel cell and hybrid vehicles in a future sustainable road transport system in the UK. *Energy Policy*, 39(4):1939–1950.

Olah, G. A. (2005). Beyond oil and gas: The methanol economy. *Angew. Chem. Int. Ed.*, 44(18):2636–2639.

Ou, X., Zhang, X., and Chang, S. (2010). Scenario analysis on alternative fuel/vehicle for China's future road transport: Life-cycle energy demand and GHG emissions. *Energy Policy*, 38(8):3943–3956.

Pachauri, R. K., Allen, M., Barros, V., Broome, J., Cramer, W., Christ, R., Church, J., Clarke, L., Dahe, Q., Dasgupta, P., et al. (2014). *Climate Change 2014: Synthesis Report. Contribution of Working Groups I, II and III to the Fifth Assessment Report of the Intergovernmental Panel on Climate Change*. IPCC.

Pal, R., Groy, T. L., and Trovitch, R. J. (2015). Conversion of carbon dioxide to methanol using a C-H activated bis(imino)pyridine molybdenum hydroboration catalyst. *Inorg. Chem.*, 54(15):7506–7515.

Pehnt, M. (2001). Life-cycle assessment of fuel cell stacks. *Int. J. Hydrogen Energy*, 26(1):91–101.

Peighambardoust, S., Rowshanzamir, S., and Amjadi, M. (2010). Review of the proton exchange membranes for fuel cell applications. *Int. J. Hydrogen Energy*, 35(17):9349–9384.

Perathoner, S. and Centi, G. (2014). CO_2 recycling: A key strategy to introduce green energy in the chemical production chain. *ChemSusChem*, 7(5):1274–1282.

Pérez-Fortes, M., Schöneberger, J. C., Boulamanti, A., Harrison, G., and Tzimas, E. (2016). Formic acid synthesis using CO_2 as raw material: Techno-economic and environmental evaluation and market potential. *Int. J. Hydrogen Energy*, 41(37):16444–16462.

Peters, M., Köhler, B., Kuckshinrichs, W., Leitner, W., Markewitz, P., and Müller, T. E. (2011). Chemical technologies for exploiting and recycling carbon dioxide into the value chain. *ChemSusChem*, 4(9):1216–1240.

Plasynski, S. I. and Ciferno, J. P. (2008). The cost of carbon dioxide capture and storage in geologic formations. Factsheet, National Energy Technology Laboratory, Washington, DC.

Rabl, A., Benoist, A., Dron, D., Peuportier, B., Spadaro, J. V., and Zoughaib, A. (2007). How to account for CO_2 emissions from biomass in an LCA. *Int. J. Life Cycle Assess.*, 12(5):281–281.

Rajagopal, D. (2017). A step towards a general framework for consequential life cycle assessment. *J. Ind. Ecol.*, 21(2):261–271.

Reiter, G. and Lindorfer, J. (2015). Global warming potential of hydrogen and methane production from renewable electricity via power-to-gas technology. *Int. J. Life Cycle Assess.*, 20(4):477–489.

REN21 (2015). *Renewables 2015 Global Status Report.* REN21, Paris.

Reuss, G., Disteldorf, W., Gamer, A. O., and Hilt, A. (2000). *Ullmann's Encyclopedia of Industrial Chemistry*, chapter Formaldehyde. Wiley-VCH Verlag GmbH & Co. KGaA.

Rihko-Struckmann, L. K., Peschel, A., Hanke-Rauschenbach, R., and Sundmacher, K. (2010). Assessment of methanol synthesis utilizing exhaust CO_2 for chemical storage of electrical energy. *Ind. Eng. Chem. Res.*, 49(21):11073–11078.

Rios, P., Curado, N., Lopez-Serrano, J., and Rodriguez, A. (2016). Selective reduction of carbon dioxide to bis(silyl)acetal catalyzed by a PBP-supported nickel complex. *Chem. Commun.*, 52(10):2114–2117.

Rönsch, S., Schneider, J., Matthischke, S., Schlüter, M., Götz, M., Lefebvre, J., Prabhakaran, P., and Bajohr, S. (2016). Review on methanation – from fundamentals to current projects. *Fuel*, 166:276–296.

Rose, L., Hussain, M., Ahmed, S., Malek, K., Costanzo, R., and Kjeang, E. (2013). A comparative life cycle assessment of diesel and compressed natural gas powered refuse collection vehicles in a Canadian city. *Energy Policy*, 52:453–461.

Schaub, T., Fries, D., Paciello, R., Mohl, K.-D., Schneider, D., and M. Schäfer, S. R. (2012). Patent CA 2774151.

Schaub, T. and Paciello, R. A. (2011). A process for the synthesis of formic acid by CO_2 hydrogenation: Thermodynamic aspects and the role of CO. *Angew. Chem. Int. Ed.*, 50(32):7278–7282.

Schiebahn, S., Grube, T., Robinius, M., Tietze, V., Kumar, B., and Stolten, D. (2015). Power to gas: Technological overview, systems analysis and economic assessment for a case study in Germany. *Int. J. Hydrogen Energy*, 40(12):4285–4294.

Schreiber, A., Zapp, P., and Kuckshinrichs, W. (2009). Environmental assessment of German electricity generation from coal-fired power plants with amine-based carbon capture. *Int. J. Life Cycle Assess.*, 14(6):547–559.

Sensfuß, F., Ragwitz, M., and Genoese, M. (2008). The merit-order effect: A detailed analysis of the price effect of renewable electricity generation on spot market prices in Germany. *Energy Policy*, 36(8):3086–3094.

Shafiee, S. and Topal, E. (2009). When will fossil fuel reserves be diminished? *Energy Policy*, 37(1):181–189.

Spath, P. and Mann, M. (2004). Life cycle assessment of renewable hydrogen production via wind/electrolysis. Report NREL/MP-560-35404, National Renewable Energy Laboratory, Washington, DC.

Spitz, P., editor (2003). *The Chemical Industry at the Millennium*, volume 978-0-941901-34-5. Chemical Heritage Foundation.

Sternberg, A. and Bardow, A. (2015). Power-to-What? - environmental assessment of energy storage systems. *Energy Environ. Sci.*, 8(2):389–400.

Sternberg, A. and Bardow, A. (2016). Life cycle assessment of Power-to-Gas: Syngas vs methane. *ACS Sustainable Chem. Eng.*, 4(8):4156–4165.

Sternberg, A., Jens, C. M., and Bardow, A. (2017). Life cycle assessment of CO_2-based C1-chemicals. *Green Chem.*, 19(9):2244–2259.

Sternberg, A., Teichgräber, H., Voll, P., and Bardow, A. (2015). CO_2 vs biomass: Identification of environmentally beneficial processes for platform chemicals from renewable carbon sources. *Comp. Aid. Chem. Eng.*, 37:1361–1366.

Steubing, B., Zah, R., and Ludwig, C. (2012). Heat, electricity, or transportation? The optimal use of residual and waste biomass in Europe from an environmental perspective. *Environ. Sci. Technol.*, 46(1):164–171.

Steward, D., Saur, G., Penev, M., and Ramsden, T. (2009). Lifecycle cost analysis of hydrogen versus other technologies for electrical energy storage. Report NREL/TP-560-46719, National Renewable Energy Laboratory, Washington, DC.

Supronowicz, W., Ignatyev, I. A., Lolli, G., Wolf, A., Zhao, L., and Mleczko, L. (2015). Formic acid: a future bridge between the power and chemical industries. *Green Chem.*, 17:2904–2911.

Thomassen, M. A., Dalgaard, R., Heijungs, R., and de Boer, I. (2008). Attributional and consequential LCA of milk production. *Int. J. Life Cycle Assess.*, 13(4):339–349.

Van-Dal, E. S. and Bouallou, C. (2013). Design and simulation of a methanol production plant from CO_2 hydrogenation. *J. Clean. Prod.*, 57:38–45.

van der Giesen, C., Kleijn, R., and Kramer, G. J. (2014). Energy and climate impacts of producing synthetic hydrocarbon fuels from CO_2. *Environ. Sci. Technol.*, 48(12):7111–7121.

Vennestrøm, P. N. R., Osmundsen, C. M., Christensen, C. H., and Taarning, E. (2011). Beyond petrochemicals: The renewable chemicals industry. *Angew. Chem. Int. Ed.*, 50(45):10502–10509.

Verdegaal, W. M., Becker, S., and Olshausen, C. v. (2015). Power-to-Liquids: Synthetisches Rohöl aus CO_2, Wasser und Sonne. *Chem. Ing. Tech.*, 87(4):340–346. (in German).

von der Assen, N. and Bardow, A. (2014). Life cycle assessment of polyols for polyurethane production using CO_2 as feedstock: Insights from an industrial case study. *Green Chem.*, 16(6):3272–3280.

von der Assen, N., Jung, J., and Bardow, A. (2013). Life-cycle assessment of carbon dioxide capture and utilization: Avoiding the pitfalls. *Energy Environ. Sci.*, 6(9):2721–2734.

von der Assen, N., Müller, L. J., Steingrube, A., Voll, P., and Bardow, A. (2016). Selecting CO_2 sources for CO_2 utilization by environmental-merit-order curves. *Environ. Sci. Technol.*, 50(3):1093–1101.

von der Assen, N., Sternberg, A., Katelhon, A., and Bardow, A. (2015). Environmental potential of carbon dioxide utilization in the polyurethane supply chain. *Faraday Discuss.*, 183:291–307.

von der Assen, N., Voll, P., Peters, M., and Bardow, A. (2014). Life cycle assessment of CO_2 capture and utilization: A tutorial review. *Chem. Soc. Rev.*, 43(23):7982–7994.

Weidema, B. (2000). Avoiding co-product allocation in life-cycle assessment. *J. Ind. Ecol.*, 4(3):11–33.

Zamagni, A., Guinée, J., Heijungs, R., Masoni, P., and Raggi, A. (2012). Lights and shadows in consequential LCA. *Int. J. Life Cycle Assess.*, 17(7):904–918.

Aachener Beiträge zur Technischen Thermodynamik

ABTT 1
Phillip Voll
Automated Optimization-Based Synthesis of Distributed Energy Supply Systems
1. Auflage 2014
ISBN 978-3-86130-474-6

ABTT 2
Johannes Jung
Comparative Life Cycle Assessment of Industrial Multi-Product Processes
1. Auflage 2014
ISBN 978-3-86130-471-5

ABTT 3
Franz Lanzerath
Modellgestützte Entwicklung von Adsobtionswärmepumpen
1. Auflage 2014
ISBN 978-3-86130-472-2

ABTT 4
Thorsten Brands
Einfluss der Gemischzusammensetzung auf die Verbrennung im Diesel- und GCAI-Motor
1. Auflage 2014
ISBN 978-3-95886-006-3

ABTT 5
Dominique Dechambre
Efficient Measurement of Liquid-Liquid Equilibria using Automation and Optimal Experimental Design
1. Auflage 2016
ISBN 978-395886-077-3

ABTT 6
Niklas von der Aßen
From Life-Cycle Assesement towards life-Cycle Design of Carbon Dioxide Capture and Utilization
1. Auflage 2016
ISBN 978-3-95886-080-3

ABTT 7
Matthias Lampe
Integrated Process and Organic Rankine Cycle Working Fluid Design in the Continuous-Molecular Targeting Framework
1. Auflage 2016
ISBN 978-3-95886-086-5

ABTT 8
Thomas Hülser
Optische Untersuchung der Zündvorgänge und deren Auswirkung auf die Verbrennung in PKW-Motoren
1. Auflage 2016
ISBN 978-3-95886-090-2

Aachener Beiträge zur Technischen Thermodynamik

ABTT 9
Malte Döntgen
Reaction Models from Reactive Molecular Dynamics and High-Level Kinetics Predictions
1. Auflage 2016
ISBN 978-3-95886-156-5

ABTT 10
Heike Schreiber
Experiments and Validated Models for Adsorption Thermal Energy Storage in Industrial and Residential Application
1. Auflage 2017
ISBN 978-3-95886-178-7

ABTT 11
André Dirk Sternberg
System-Wide Perspective for Life Cycle Assesment of CO_2-based C1-Chemicals
1. Auflage 2017
ISBN 978-3-95886-193-0